Madalina Xenia Calbureanu Popescu
Raluca Anda Malciu

Study of Vibrations Influence upon the Response of Different Materials

AF153077

Madalina Xenia Calbureanu Popescu
Raluca Anda Malciu

Study of Vibrations Influence upon the Response of Different Materials

LAP LAMBERT Academic Publishing

Impressum / Imprint

Bibliografische Information der Deutschen Nationalbibliothek: Die Deutsche Nationalbibliothek verzeichnet diese Publikation in der Deutschen Nationalbibliografie; detaillierte bibliografische Daten sind im Internet über http://dnb.d-nb.de abrufbar.

Alle in diesem Buch genannten Marken und Produktnamen unterliegen warenzeichen-, marken- oder patentrechtlichem Schutz bzw. sind Warenzeichen oder eingetragene Warenzeichen der jeweiligen Inhaber. Die Wiedergabe von Marken, Produktnamen, Gebrauchsnamen, Handelsnamen, Warenbezeichnungen u.s.w. in diesem Werk berechtigt auch ohne besondere Kennzeichnung nicht zu der Annahme, dass solche Namen im Sinne der Warenzeichen- und Markenschutzgesetzgebung als frei zu betrachten wären und daher von jedermann benutzt werden dürften.

Bibliographic information published by the Deutsche Nationalbibliothek: The Deutsche Nationalbibliothek lists this publication in the Deutsche Nationalbibliografie; detailed bibliographic data are available in the Internet at http://dnb.d-nb.de.

Any brand names and product names mentioned in this book are subject to trademark, brand or patent protection and are trademarks or registered trademarks of their respective holders. The use of brand names, product names, common names, trade names, product descriptions etc. even without a particular marking in this works is in no way to be construed to mean that such names may be regarded as unrestricted in respect of trademark and brand protection legislation and could thus be used by anyone.

Coverbild / Cover image: www.ingimage.com

Verlag / Publisher:
LAP LAMBERT Academic Publishing
ist ein Imprint der / is a trademark of
OmniScriptum GmbH & Co. KG
Heinrich-Böcking-Str. 6-8, 66121 Saarbrücken, Deutschland / Germany
Email: info@lap-publishing.com

Herstellung: siehe letzte Seite /
Printed at: see last page
ISBN: 978-3-8473-2330-3

THE STUDY OF VIBRATIONS INFLUENCE UPON THE RESPONSE OF DIFFERENT KINDS OF MATERIALS - VISCOUSELASTIC AND ELASTIC- FOR BAR-TYPE LINKS

Associate Professor PhD Eng. Madalina Calbureanu
PhD Eng. Raluca Anda Malciu

"Science never solves a problem without creating ten more."
George Bernard Shaw

Summary

Why this research?

Research of recent years regarding bar-type links vibrations gained an increasingly theoretical and applied importance because knowing the vibrations propagation and the dynamic effects of strains produced during working depending on loads, geometric and mechanical features of bars led to the finding of advantageous technical solutions in design and implementation of mechanisms which have to operate at high speeds or mechanisms whose links positions have to be accurate.

In research carried out in last years, dynamic analysis of mechanisms was based on the assumption that kinematic elements behave as rigid solids. As a result of increasingly stringent requirements in terms of increasingly higher speed and accuracy of kinematic elements position, it is necessary to take into account the dynamic effects due to elastic deformation occurring during mechanisms operation.

Mechanisms having as component kinematic elements gears, translational pair and so on are not rigid in reality – they are elastic and they deform subjected to high static or dynamic forces. Moving at low speed, when the static forces aren't significant, it is not necessary to take into account the elastic deformations, but in the applications with high speed, mechanisms may be designed without being operable in reality due to the large fluctuations of inertial forces. An important role in errors occurring during the operation of mechanisms with elastic kinematic elements belongs to the influence of the lubricant film formed in translational pairs exerted on elastic links vibrations.

This monograph used in experimental investigations a crank and connecting rod mechanism with one of the kinematic links made of different materials. The longest link was made of linear viscoelastic material in the first part of the study and the stress and strains were highlighted; the influence of the behavior of this particular type of material above the vibrations of the kinematic element was studied. The second part of the study used a deformable elastic material for the longest link. Moreover this link was subjected to the pressure of the oil from a rectilinear pair during the motion of the mechanism. There were used in experimental investigations different types of oils as lubricants of the rectilinear pair and the vibrations of the elastic kinematic element were studied.

A lot of instruments were used in this research such as FEA (COSMOSM), CAD (NASTRAN, SOLIDWORKS) programs, MAPLE and MATHLAB.

1. SOME IMPORTANT THINGS ABOUT THE VISCOELASTIC VS ELASTIC MATERIALS BEHAVIOUR IN DYNAMIC ANALYSIS

> "Intelligence is the ability to adapt to change."
> **Stephen Hawking**

1.1 INTRODUCTION

Research of recent years regarding bar-type links vibrations gained an increasingly theoretical and applied importance because knowing the vibrations propagation and the dynamic effects of strains produced during working depending on loads, geometric and mechanical features of bars led to the finding of advantageous technical solutions in design and implementation of mechanisms which have to operate at high speeds or mechanisms whose links positions have to be accurate.

1.1.1 MATHEMATICAL MODELS IN DISPLACEMENTS FOR THE VIBRATIONS OF BAR-TYPE LINKS WITH LINEAR ELASTIC BEHAVIOUR

In order to obtain the mathematical models in displacements for the vibrations of bar-type links with linear elastic behavior, it was started from the equations of motion in displacements, for a straight link, linearly elastic, in roto-translation motion, where the total mass m_t is considered concentrated in "n" points and the method of the influence coefficients was applied.

$$\sum_{k=1}^{n} \beta_{ik} m_k \left[\ddot{u}_{1k} - 2\omega\dot{u}_{2k} - \omega^2 u_{1k} - \varepsilon u_{2k} + a_{01} - \omega^2 x_k \right] + Eu_{1i} = \sum_{k=1}^{n} \beta_{ik} F_{1k}; \quad i = \overline{1,n}; (1.1)$$

$$\sum_{k=1}^{n} \alpha_{ik} m_k \left[\ddot{u}_{2k} + 2\omega\dot{u}_{1k} - \omega^2 u_{2k} + \varepsilon u_{1k} + a_{02} + \varepsilon x_k \right] + Eu_{2i} = \sum_{k=1}^{n} (\alpha_{ik} F_{2k} + \delta_{ik} M_{3k}); \quad i = \overline{1,n};$$

$$(1.2)$$

where:

- $\dfrac{\beta_{ik}}{E}, i, k = \overline{1,n}$ -the influence coefficients for the longitudinal vibrations $u_{1i}(t)$;

- $\dfrac{\alpha_{ik}}{E}; \dfrac{\delta_{ik}}{E}, i, k = \overline{1,n}$ -the influence coefficients for the transverse vibrations $u_{2i}(t)$;

For a straight, linearly elastic link, model of continuous medium in planar roto-translation motion, the equations of motion in displacements, in the coupled version, were obtained using Hamilton's principle from elasto-dynamics, by neglecting the cutting forces influence.

Written in matrix formalism, the above equations looks like this:

$$[\beta][m](\{\ddot{u}_1\} - 2\omega\{\dot{u}_2\} - \omega^2\{u_1\} - \varepsilon\{u_2\} + \{A_1\}) + E\{u_1\} = [\beta]\{F_1\}; \qquad (1.3)$$

$$[\alpha][m](\{\ddot{u}_2\} + 2\omega\{\dot{u}_1\} - \omega^2\{u_2\} + \varepsilon_3\{u_1\} + \{A_2\}) + E\{u_2\} = [\alpha]\{F_2\} + [\delta]\{M_3\}. \qquad (1.4)$$

The mathematical models found for the planar roto-translation motions can be written using two differential matrix form operators $[L_0]$ and $[L_1]$, by grouping the terms of coupling between the longitudinal and transverse vibrations and the terms which gives the quality of models varying in time.

$$[L_0]\{u\} + [M_4]\{a_0\} + \{V_1\} + [M_7]\{f\} + \{V_2\} + [L_1]\{u\} = \{0\}, \qquad (1.5)$$

where:
$$[L_0]\cdot = [M_1]\frac{\partial^4\cdot}{\partial x^4} + [M_2]\frac{\partial^4\cdot}{\partial x^2\partial t^2} + [M_3]\frac{\partial^2\cdot}{\partial x^2} + [M_4]\frac{\partial^2\cdot}{\partial t^2}; \qquad (1.6)$$

$$[L_1]\cdot = [M_5]\frac{\partial\cdot}{\partial t} + [M_6]\cdot, \qquad (1.7)$$

By neglecting the term which includes differential matrix form operator $[L_1]$, it is obtained the decoupled, linear model with constant coefficients, in the first approximation:

$$[L_0]\{u\} + [M_4]\{a_0\} + \{V_1\} + [M_7]\{f\} + \{V_2\} = \{0\}. \qquad (1.8)$$

1.1.2 MATHEMATICAL MODELS IN DISPLACEMENTS FOR THE VIBRATIONS OF BAR-TYPE LINKS WITH LINEAR VISCOELASTIC BEHAVIOUR

Mathematical models in displacements for the vibrations of straight bar-type links with linear viscoelastic behavior for mechanical models with a finite number of degrees of freedom were obtained from classical linear equations of elasto-dynamics. Based on elasto-viscoelastic analogies stated by Alfrey and Lee, it was applied the unilateral Laplace transform with respect to time to these equations. In the case of equations of motion obtained with Hamilton's principle, it was applied the same method, substituting Young's modulus of elasticity E with its Laplace transform $\tilde{E}(s)$.

6

For simple rheological models such as Kelvin-Voight, Standard, Maxwell, Zener or Hooke model tied in parallel with Maxwell element, $\tilde{E}(s)$ is determined by the relationship:

$$\tilde{E}(s) = \frac{a_0 s + a_1}{b_0 s + b_1}, \tag{1.9}$$

and, for the more complicated models, such as, for example, the Burgers model, the relationship becomes:

$$\tilde{E}(s) = \frac{A_0 s^2 + a_0 s + a_1}{B_0 s^2 + b_0 s + b_1}, \tag{1.10}$$

where the coefficients A_0, B_0, a_0, b_0, a_1 and b_1 are determined according to [1].

Inverting the Laplace transform in equation (1.9), we obtain Young's modulus as a function of time as follows:

$$E(t) = k_1 \delta(t) + k_2 e^{-k_3 t}, \tag{1.11}$$

where: $k_1 = \dfrac{a_0}{b_0}$; $\delta(t)$ - Dirac function; $k_2 = \dfrac{a_1 b_0 - a_0 b_1}{b_0^2}$; $k_3 = \dfrac{b_1}{b_0}$.

Applying the inverse Laplace transform to relation (1.10), we obtain Young's modulus as a function of time, as follows:

$$E(t) = k_1 \delta(t) + k_2 e^{-k_3 t} [k_4 \sin(k_5 t) + k_6 \cos(k_5 t)], \tag{1.12}$$

where: $k_1 = \dfrac{A_0}{B_0}$; $k_2 = \dfrac{1}{B_0^2 \sqrt{4 B_0 b_1 - b_0^2}}$; $k_3 = \dfrac{b_0}{2 B_0}$;

$$k_4 = 2 B_0 (B_0 a_1 - A_0 b_1) - b_0 (B_0 a_0 - A_0 b_0); \quad k_5 = \frac{\sqrt{4 B_0 b_1 - b_0^2}}{2 B_0}; \tag{1.13}$$

$$k_6 = (B_0 a_0 - A_0 b_0) \sqrt{4 B_0 b_1 - b_0^2}.$$

In order to get the mathematical models in real time for continuous media in planar motions, there were deliberately neglected the terms that don't allow the direct application of the Laplace transform with respect to time, i.e. the coupling terms and those terms having coefficients as functions of time. So, it was obtain the mathematical model with equations with partial derivatives for the first approximation, which is solved using the method of successive approximations.

The matrix equation of the first approximation in Laplace images, for the vibration of a viscoelastic bar-type link in a plane rototranslation motion:

$$[\tilde{L}_0(s)]\{\tilde{u}^{(1)}\}+[M_4]\{\tilde{a}_0\}+\{\tilde{v}_1\}+[M_7]\{\tilde{f}\}+\{\tilde{v}_2\}+[M_2]\frac{d^2\{V_3\}}{dx^2}+[M_4]\{V_3\}=\{0\}$$

(1.14)

where: $[\tilde{L}_0(s)]\cdot=[M_1(s)]\frac{\partial^4\cdot}{\partial x^4}+[M_8(s)]\frac{\partial^4\cdot}{\partial x^2}+[M_4(s)]\cdot$

The involved matrices and vectors have the form given by [1].

1.1.3 METHODS FOR SOLVING THE MATHEMATICAL MODELS FOR THE VIBRATIONS OF BAR-TYPE LINKS WITH LINEAR ELASTIC BEHAVIOUR

Solving the mathematical models presented for straight bars-type links with linear elastic behavior was done using Laplace transform and Fourier transform finite in sine or cosine in the terms of boundary conditions specific for technical applications. There were obtained algebraic systems where the unknowns were the displacements in their Laplace and Fourier images. The longitudinal and transversal displacements fields for bar-type links with linear elastic behavior were obtained by reversing the integral transforms.

For mechanical models with a finite number of degrees of freedom, Laplace transform with respect to time was applied to the system of differential equations representing the mathematical model of the motion. This gave an algebraic system of equations, the unknowns being the Laplace images of the displacements, system which was solved elementary. The displacements fields were obtained by reversing the Laplace transform.

For the mathematical model of the first approximation for free vibrations, which was a decoupled model, linear and with constant coefficients, it was applied the unilateral Laplace transform with respect to time and then, to the first equation it was applied the Fourier transform in finite cosine and to the second equation the Fourier transform in finite sinus. It was obtained a decoupled algebraic system, where the unknowns were the displacements in their Laplace and Fourier images in the cosine, respectively in the sine.

It was taken into account the boundary conditions which allowed the application of the two Fourier transforms to the original functions and to their Laplace images respectively. Then, there were reversed the Laplace and Fourier transforms and it was obtained the solution in the first approximation $\{u^{(1)}(x,t)\}$.

Now the vector $[L_1]\{u^{(1)}\}$ may be determined in a first approximation with the found $\{u^{(1)}(x,t)\}$ and, if it is introduced in the general equation, the mathematical model in the second approximation is obtained. Its resolution with the help of the integral transforms gives the solution in the second approximation. The iterative process continues and the mathematical model in the "j"-th approximation is obtained.

$$[L_0]\{u^{(j)}(x,t)\}+[M_4]\{a_0\}+\{V_1\}+[M_7]\{f\}+\{V_2\}+[L_1]\{u^{(j-1)}(x,t)\}=\{0\} \qquad (1.15)$$

where j = 1, 2,..., n and $[L_1]\{u^{(0)}\}=\{0\}$.

The solution in the "j"-th approximation (1.16), (1.17) is obtained in the same way and the process of successive approximations continue until the difference between two consecutive solutions is less than $\varepsilon > 0$ and sufficiently small, depending on the required accuracy of the calculus (1.18).

$$u_1^{(j)}(x,t)=\frac{1}{L}\cdot u_{1,c}^{(j)}(0,t)+\frac{2}{L}\sum_{n=1}^{n=\infty}u_{1,c}^{(j)}(n,t)\cdot\cos(\alpha_n\cdot x) \qquad (1.16)$$

$$u_2^{(j)}(x,t)=\frac{2}{L}\sum_{n=1}^{n=\infty}u_{2,s}^{(j)}(n,t)\cdot\sin(\alpha_n\cdot x), \qquad (1.17)$$

where $u_{1,c}^{(j)}(n,t)$ is the integral Fourier transform in finite cosine of the longitudinal elastic displacement and $u_{2,s}^{(j)}(n,t)$ is the integral Fourier transform in finite sine of the transverse elastic displacement.

The process of successive approximations continues until:

$$\forall n, \left\|\{u^{(j)}\}-\{u^{(j-1)}\}\right\| \leq \varepsilon, \qquad (1.18)$$

where: - $\{u^{(j)}\}=\{u_1^{(j)}(x,t),u_2^{(j)}(x,t)\}^T$;

$\qquad$ - $\{u^{(j-1)}\}=\{u_1^{(j-1)}(x,t),u_2^{(j-1)}(x,t)\}^T$.

1.1.4 METHODS FOR SOLVING THE MATHEMATICAL MODELS FOR THE VIBRATIONS OF BAR-TYPE LINKS WITH LINEAR VISCOELASTIC BEHAVIOUR

The mathematical models for straight bar-type links with linear viscoelastic behavior were solved using the accurate and the iterative methods, by applying to them the integral Fourier transforms in finite sine or cosine. Then, there were solved the resulted algebraic system of equations and the unknowns $\tilde{u}_i^*(n,s)$ where $i = 1,...,3$, for spatial motion and $i = 1, 2$, for planar motion, were determined as:

$$\tilde{u}_{i,s}^*(n,s) = \frac{P_{i,s}(n,s)}{Q_i(n,s)} \quad \text{or} \quad \tilde{u}_{i,c}^*(n,s) = \frac{P_{i,c}(n,s)}{Q_i(n,s)}, \quad (1.19)$$

where $i = 1,...,3$, for spatial motion and $i = 1, 2$, for planar motion; the polynomials $Q_i(n,s)$ and the functions $P_{i,s}(n,s)$, respectively $P_{i,c}(n,s)$, are determined for each type of motion.

The solutions in Laplace images were obtained by reversing the Fourier transforms.

$$\tilde{u}_i(x,s) = \frac{2}{L}\sum_{n=1}^{\infty}\tilde{u}_{i,s}^*(n,s)\sin(\alpha_n x)$$

$$\text{or} \quad \tilde{u}_i(x,s) = \frac{1}{L}\tilde{u}_{i,c}^*(0,s) + \frac{2}{L}\sum_{n=1}^{\infty}\tilde{u}_{i,c}^*(n,s)\cos(\alpha_n x), \quad (1.20)$$

where $\alpha_n = \dfrac{n\pi}{L}$.

The solutions $u_i(x, t)$, giving the longitudinal and transversal displacements were obtained by reversing the Laplace transform with the help of the development theorems and numerical methods.

So, in order to solve the mathematical model in displacements in "real time" for the vibrations of bar-type and viscoelastic link, there were determined the polynomial expressions $Q_i(n,s)$ and the functions $P_{i,s}(n,s)$ and $P_{i,c}(n,s)$ met in the relations (1.19).

Solving of the mathematical models obtained in real-time by iterative methods involved first the solving of matrix equations that gave the models in the first approximation, determining the solution in the first approximation $\{u^{(1)}(x,t)\}$, using finite Fourier transforms in sine and cosine, for simplified boundary conditions. Then it was calculate the term $[L_1]\{u^{(1)}\}$ of the equation (1.5) in the first approximation and it was determined the mathematical model of the second approximation.

The new model was solved using the integral transformation, having as result the solution $\{u^{(2)}(x,t)\}$ in the second approximation. Continuing the same way, it was obtained a mathematical model for the "j" –th approximation as:

$$[L_0]\{u^{(j)}(x,t)\} + [M_4]\{a_0\} + \{V_1\} + [M_7]\{f\} + \{V_2\} + [L_1]\{u^{(j-1)}(x,t)\} = \{0\} \quad (1.21)$$

where:

$$[L_1]\{u^{(j-1)}(x,t)\}=[M_5]\frac{\partial\{u^{(j-1)}(x,t)\}}{\partial t}+[M_6]\{u^{(j-1)}(x,t)\}, \quad j=\overline{1,n}, \quad (1.22)$$

with $[L_1]\{u^{(0)}\}=\{0\}$.

The process of successive approximations continued until:

$$\left|\{u^{(j)}-u^{(j-1)}\}\right|\le\varepsilon, \forall n, \forall t, \qquad (1.23)$$

where $\varepsilon>0$ had a size order corresponding to the required calculation accuracy.

Inverting the Fourier transformation, the obtained solutions in Laplace images were:

$$\tilde{u}_i^{(1)}(x,s)=\frac{2}{L}\sum_{n=1}^{\infty}\tilde{u}_{i,s}^{*(1)}(n,s)\sin(\alpha_n x)$$

or

$$\tilde{u}_i^{(1)}(x,s)=\frac{1}{L}\tilde{u}_{i,c}^{*(1)}(0,s)+\frac{2}{L}\sum_{n=1}^{\infty}\tilde{u}_{i,c}^{*(1)}(n,s)\cos(\alpha_n x), \qquad (1.24)$$

where $\alpha_n=\dfrac{n\pi}{L}$.

Inverting the Laplace transforms with the help of the developing theorems and numerical methods for solving the equations $Q_i(n, s) = 0$, the obtained solutions $u_i^{(1)}(x,t)$ gave longitudinal and transverse displacement field in the first approximation.

Then there were determined the term $[L_1]\{u^{(1)}\}$, the mathematical model of the second approximation and the solution $\{u^{(2)}(x,t)\}$ in the second approximation, and so on, it was determined the mathematical model for the "j"-th approximation using the relations (1.21) and (1.22).

After applying the Fourier integral transformation to this model, there were determined the solutions for the "j"-th approximation:

$$\tilde{u}_{i,s}^{*(k)}(n,s)=\frac{P_{i,s}^{(1)}(n,s)\pm\alpha_n(b_0 s+b_1)[L_1]\{u_{i,s}^{*(k-1)}(n,s)\}}{Q_i(n,s)} \qquad (1.25)$$

or

$$\tilde{u}_{i,c}^{*(k)}(n,s)=\frac{P_{i,c}^{(1)}(n,s)\pm\alpha_n(b_0 s+b_1)[L_1]\{u_{i,c}^{*(k-1)}(n,s)\}}{Q_i(n,s)}, \qquad (1.26)$$

where i = 1, 2, the sign "+" was considered for i = 1, and the sign "-" for i = 2.

Reversing the finite Fourier transforms, we obtained the solutions in Laplace images of the kind given by relations (1.24) and finally, reversing Laplace transforms with developing theorems and numerical methods, there were obtained the solutions for the longitudinal and transverse displacement field.

11

The solutions determined as shown above highlighted the effect of the motion kinematic parameters on the vibration of certain machines kinematic chains, kinematic chains composed of bar-type links with viscoelastic behavior.

Determination of displacements due to vibration, depending on the kinematic parameters of the motion is very important in the activity of construction design of these kinematic chains because it makes possible the determination of the components of the stress tensor and deformation tensor. These elements are necessary for dimensioning and verification computations specific to the design activity.

> "Research is what I'm doing when I don't know what I'm doing."
> **Wernher von Braun**

1.2 VIBRATIONS INFLUENCE UPON THE STRESS AND STRAINS STATES OF BAR-TYPE LINKS

1.2.1 THEORETICAL RESULTS

Referring to the bars with linear elastic behavior, it is known that a state of elastic body is completely determined by the stress tensor $T_\sigma = (\sigma_{ij})$, the strains tensor $T_\varepsilon = (\varepsilon_{ij})$, $i, j = 1 \div 3$, and the displacement vector $\bar{u}$. These three elements contain a total of 15 unknowns, which are functions of spatial coordinates x_i from elastostatics. The 15 unknowns are linked together by 3 independent groups of equations, a total of 15 equations, too:

- Equations of equilibrium:

$$\sigma_{ij,j} + f_i = 0; \quad i,j = \overline{1,3}, \tag{1.27}$$

where f_i are the projections on coordinate axes of the volumetric density of strengths acting upon elastic body;

- Equations of Cauchy (geometric equations):

$$\varepsilon_{ij} = \frac{1}{2}\left(\frac{\partial u_i}{\partial x_j} + \frac{\partial u_j}{\partial x_i}\right); \quad i,j = \overline{1,3}; \tag{1.28}$$

- Hooke's constitution law for the isotropic elastic body (physical equations):

$$\sigma_{ij} = \lambda \varepsilon \delta_{ij} + 2\mu \varepsilon_{ij}; \quad i,j = \overline{1,3}, \tag{1.29}$$

where ε -specific volumetric deformation: $\varepsilon = \varepsilon_{11} + \varepsilon_{22} + \varepsilon_{33}$, λ, μ - the Lamé's parameter, $\delta_{ij} = \begin{cases} 1; i = j; \\ 0; i \neq j; \end{cases}$ $i, j = \overline{1,3}$ - Kronecker's tensor.

So, for linear elastic links subjected to vibrations, the determination of displacements fields depending on the kinematic parameters of the motion makes possible the calculus of the components of the additional strains tensor that occur due to vibrations, and then calculate the components of the additional stress tensor. Thus it can be made evident the influence of vibrations upon stress and strains states that arise in a link of a mechanism during its operation, obtaining the required data for rigorous constructive designing of the mechanism bar-type components.

For example it was considered the linear elastic connecting rod of a R(RRT) mechanism, for which the longitudinal and transversal displacements fields were computed by the iterative method presented in 1.1 paragraph.

Considering the mathematical model for coupled equations in the case of free vibration, it was obtained the decoupled and linear model with constant coefficients in the first approximation:

$$[L_0]\{u\} + [M_4]\{a_0\} + \{V_1\} = \{0\}, \tag{1.30}$$

where:

$$\{a_0\} = \left\{ \omega_0^2 r \left[\frac{r}{L} \sin^2(\omega_0 t) - \cos(\omega_0 t) \right] - \omega_0^2 r \left[\sin(\omega_0 t) + \frac{r}{2L} \sin(2\omega_0 t) \right] \right\}^T, \tag{1.31}$$

$$\omega = -\frac{r}{L} \omega_0 \cos(\omega_0 t); \varepsilon = \frac{\omega_0^2 r}{L} \sin(\omega_0 t). \tag{1.32}$$

It was applied the unilateral Laplace transform with respect to time to the equation (1.30) and Young's modulus, E, was replaced with $\tilde{E}(s)$. We obtained the first matrix equation of the first approximation in Laplace images, for the vibration of the R (RRT) mechanism viscoelastic rod as follows:

$$[L_0(s)]\{\tilde{u}^{(1)}\} + [M_4]\{\tilde{a}_0\} + \{\tilde{V}_1\} = \{0\}, \tag{1.33}$$

For $\tilde{E}(s)$ it was considered the appropriate expression corresponding to the Maxwell mechanical model, specific to polymers in solid state (plastics):

$$\tilde{E}(s) = \frac{a_0 s}{b_0 s + b_1} \tag{1.34}$$

where, according to [1], we had:

$$a_0 = 9GK; b_0 = 3K + G; b_1 = \frac{3KG}{\eta};$$

(1.35)

G - transverse elasticity modulus;

K - modulus of compressibility:

$$K = \frac{\nu E}{(1+\nu)(1-2\nu)} + \frac{2}{3}G;$$

(1.36)

ν - Poisson's coefficient of transverse contraction;

η - a constant corresponding to the Newtonian component of Maxwell model.

The finite Fourier transformation in cosine was applied to the first equation of the system (1.33) and finite Fourier transformation in sine was applied to the second equation. Coupled algebraic systems were obtained, where the unknowns were represented by the displacements in Laplace and Fourier images in cosine, $u_{1,c}^{(1)}(n,s)$ and, respectively, in sine, $u_{2,s}^{(1)}(n,s)$.

There were taken into account the boundary conditions and the initial conditions, which allowed the application of the second Fourier transform to the original functions, respectively, to their Laplace images.

Then, there were inverted the Laplace and Fourier transforms and the solution in the first approximation $\{u^{(1)}(x,t)\}$ was obtained. Next, it was determined the term $[L_1]\{u^{(1)}\}$ of equation (1.5) in the first approximation and the mathematical model of the second approximation, according to the algorithm presented in Section 1.1.4.

The new model was solved using the integral transformations, obtaining as result the solution in the second approximation $\{u^{(2)}(x,t)\}$. Continuing the same way, we obtained the mathematical model for the "j"-th approximation, having the expression given by (1.21).

His solution was the solution the "j"-th approximation given by the expression:

$$u_1^{(j)}(x,t) = \frac{1}{L} \cdot u_{1,c}^{(j)}(o,t) + \frac{2}{L}\sum_{n=1}^{n=\infty} u_{1,c}^{(j)}(n,t) \cdot \cos(\alpha_n \cdot x) ,$$

(1.37)

$$u_2^{(j)}(x,t) = \frac{2}{L}\sum_{n=1}^{n=\infty} u_{2,s}^{(j)}(n,t) \cdot \sin(\alpha_n \cdot x) ,$$

(1.38)

where $u_{1,c}^{(j)}(n,t)$ is the finite Fourier transform in cosine of the longitudinal elastic displacement and $u_{2,s}^{(j)}(n,t)$ is the finite Fourier transform in sine of the elastic transverse displacement.

In the case of bars with linear viscoelastic behavior, it is known that linear viscoelastic bodies are comprised of two different environments, one with the properties of elastic body and one with properties of viscous fluid. In these bodies there are found instant strains growing limited or unlimited in time, a phenomenon called *creep*, and variations of tension in relation to time in the body, by maintaining a constant strain and temperature, a phenomenon called *relaxation*.

Viscoelastic solids are characterized by their ability to accumulate and distribute mechanical energy; they are part of the class of bodies with memory, their current stress state depending on the history of the suffered strains.

The complete system of equations for viscoelastic solids is written in a similar form as the one of linear elastic solid, with the difference that the functions that occur are distribution-type functions in the distributions field D'$_+$, depending on time t $\in$ R and on the parameter r$\in$ $\Omega \subset$ R^3 and with discontinuities of first kind in origin.

It was demonstrated that the constitutive laws of the two models of solids are in a strong dependence consisting of the fact that the Laplace or Fourier's images in distributions of the constitutive law of the solid viscoelastic coincide with the mathematical structure of the corresponding Hooke's law for the elastic solid, the complex variable "s" of the Laplace transform having the role of parameter r $\in$ $\Omega \subset$ R^3. This fact led Alfrey and Lee to the formulation of the *correspondence principle*, which may sound like this: in order to solve a viscoelasticity problem the correspondent problem from elasto-dynamics has to be solved and the Laplace image of the obtained solution has to be considered, the constants being replaced by the Laplace images of the corresponding quantities from viscoelasticity.

In technical literature it can be found determination results for the characteristics of materials with viscoelastic behavior obtained under conditions of SR ISO 178 from 1998 relative to the determination of bending characteristics of rigid plastics using static bending test at standard temperature.

The apparent modulus of elasticity for bending as approximate value of Young's modulus, required in determining the shear modulus G and the bulk modulus K, was determined for polyvinyl chloride, using the initial linear portion of stress-strain curves. Also, the constant η of the related Newtonian component of Maxwell rheological model was determined depending on the stress at the maximum bending load σ_0 and the inclination of the line representing the area of stabilized creep. So, the Laplace transform of Young's modulus $\tilde{E}(s)$.was determined with the found values of the constants G, K and η.

It can be concluded that the displacements fields can be determined depending on the kinematical parameters of motion for viscoelastic links, too. Then, the components of the additional strains tensor that occur due to vibrations can be computed depending on the vibrations displacements fields and further, the components of the additional stress tensor. Thus it is emphasizing the influence of vibrations on the states of strains and stress that appear in a link of a mechanism, during its action. This makes possible to solve problems of links correctly sizing, eliminating the possibility of over-sizing them or their defective exploitation.

For the connecting rod of the R(RRT)mechanism, made from a viscoelastic material, the components of the strains tensor and of the stress tensor are determined analytically using the displacements fields previously calculated.

"Computers are magnificent tools for the realization of our dreams, but no machine can replace the human spark of spirit, [..], and understanding."
Louis Gerstner

1.2.2 NUMERICAL APPLICATIONS

The methods described above are applied to the vibrations calculus for a connecting rod, part of a crank and connecting rod mechanism, when it is made of steel and when the connecting rod is made of un-masticated polyvinyl chloride, PVC-U (EN ISO 12608: 2003).

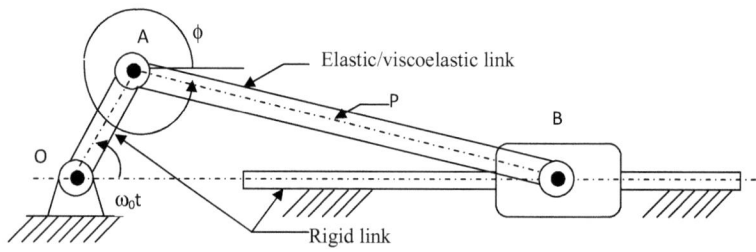

Fig. 1.1 The crank and connecting rod mechanism

Polyvinyl chloride is one of the standard plastics today, with polyethylene (PE), polypropylene (PP) and polystynol (PS). Obtained from carbon, hydrogen and chlorine, polyvinyl chloride, thanks to its miscibility with various additives for improving its quality, gain a wide range of properties, which cause a wide range of specific products.

The longitudinal displacements $u_1(x,t)$ and the transversal displacements $u_2(x,t)$ due to free vibrations in the first approximation were computed using the relations from paragraph 1.1.3 and Mathematica program, for the connecting rod made of steel and a driving link speed of 206.5 rpm.

In Fig. 1.2 there are presented the time variation diagrams of linear elastic displacements of the driving link for a rotation speed n = 206.5 **rpm** and in Fig. 1.3 the time variation of these displacements for a point placed at 1/4 of the connecting rod length from the end of action.

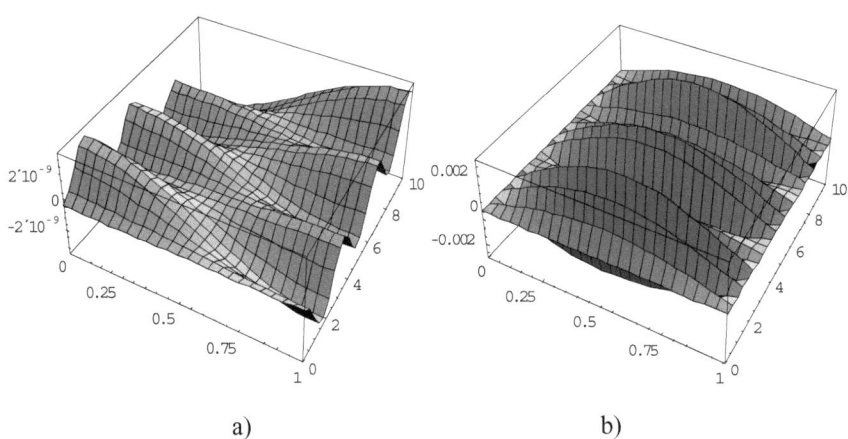

a)   b)

Fig. 1.2 Time variation diagrams of linear elastic displacements of the driving link for a rotation speed n = 206.5 **rpm**:

a) longitudinal displacements $u_1^{(1)}(x,t)$; b) transverse displacements $u_2^{(1)}(x,t)$

The variations of vibration accelerations in the mentioned point are obtained as the second derivative in relation to time of the vibrations longitudinal displacement, $u_1(L/4,t)$, and vibrations transversal displacement, $u_2(L/4,t)$. The variation diagrams of the vibration accelerations are determined in the described manner for certain speeds of the driving link and certain locations of a point on the connecting rod made of steel, according to the performed experimental measurements. In Fig. 1.4 there are presented the time variation of vibration accelerations for a point placed at 1/4 of the connecting rod length from the end of action.

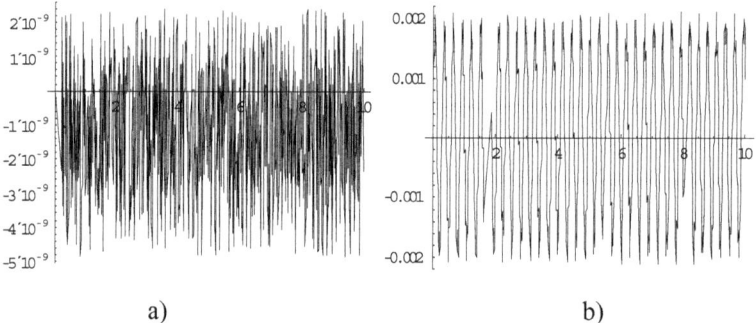

a)   b)

Fig. 1.3 Time variation diagrams for the linear elastic displacements of a point placed at 1/4 of the connecting rod length to the point A, for the rotation speed n = 206.5 rpm: a) longitudinal displacements, $u_1^{(1)}(L/4,t)$; b) transverse displacements $u_2^{(1)}(L/4,t)$.

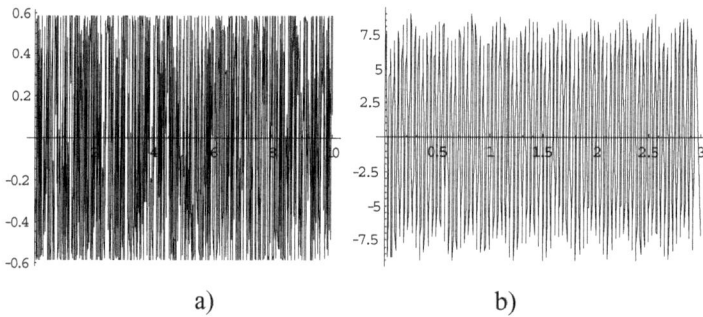

a)   b)

Fig. 1.4 Time variation diagrams of the vibration accelerations of the point placed at ¼ of the connecting rod length to the point A, for the rotation speed n = 206.5 rpm: a) longitudinal vibration acceleration a_1 (L/4,t); b) transverse vibration acceleration, a_2(L/4,t).

The accelerations effective values for the transverse vibrations were determined for the entire range of speed of the driving link used in the experimental determinations with the connecting rod made of steel, based on the vibration accelerations calculus with Mathematica program and on their variation diagrams depending on the speed of the driving link of the type shown, in order to compare the computed values with the measured ones. In Table 1.1 there are presented the effective values of vibration acceleration in transverse-vertical direction, measured for the connecting rod made of steel.

It was noticed that the calculated (theoretical) values of the vibration acceleration on the vertical direction increased with an increasing action frequency or speed of the driving link and the highest values were recorded in the point placed at 1/4 of the length of the connecting rod measured from the end of action (see Fig. 1.5).

Table 1.1 - Effective vibration acceleration of the connecting rod made of steel in transverse-vertical direction

No.	Link material/Place of measurement/Frequency (Hz)	Effective theoretical value of vibration acceleration in the vertical direction, $a_2(x,t)$ (m/s^2)
1	Steel/1_4/1.758	1.9692
2	Steel /1_4/2.564	2.9909
3	Steel /1_4/3.442	6.1050
4	Steel /1_4/3.955	9.0765
5	Steel /1_2/1.758	1.4142
6	Steel /1_2/2.490	1.7703
7	Steel /1_2/3.662	4.0369
8	Steel /1_2/4.614	8.3743
9	Steel /3_4/1.538	1.2423
1	Steel /3_4/2.564	1.5552
1	Steel /3_4/3.735	3.7666
1	Steel /3_4/4.321	5.4668

Then, the longitudinal displacements $u_1(x,t)$ and the transversal displacements $u_2(x,t)$ in the first approximation were calculated with Mathematica program for the connecting rod made of un-masticated polyvinyl chloride for different speeds between 100 and 285 rpm, and the correspondent variations in time diagrams were obtained. The variation of these displacements was determined in a point placed at 1/4 of the length of the connecting rod measured from the end of action.

The variations of vibration accelerations in the mentioned point were obtained as the second derivative in relation to time of the vibrations longitudinal displacement, $u_1(L/4,t)$, and vibrations transversal displacement, $u_2(L/4,t)$.

The accelerations effective values for the transverse vibrations were determined for the entire range of speed of the driving link used in the experimental determinations with the connecting rod made of un-masticated polyvinyl chloride, based on the vibration accelerations calculus with Mathematica program and on their variation diagrams depending on the speed of the driving link of the type shown, in order to compare the computed values with the measured ones.

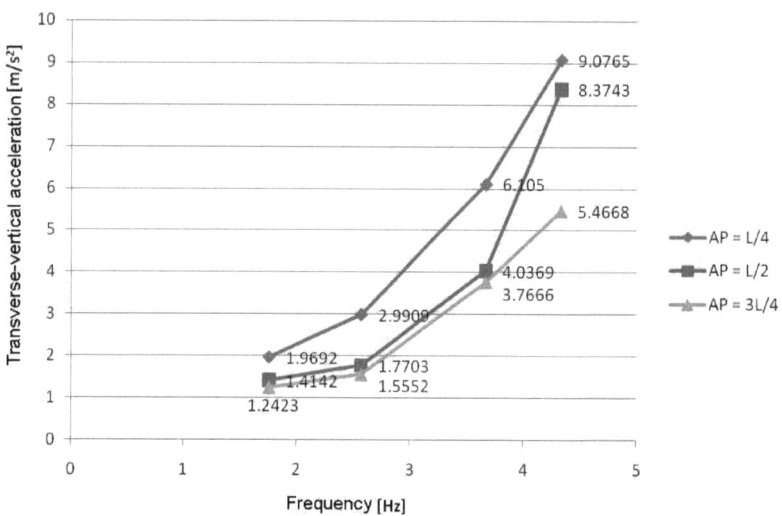

Fig. 1.5 Variation of transverse-vertical acceleration of a point P placed on the steel connecting rod depending on the driving frequency for three considered positions of the point P (AP = L/4, AP =L/2 and AP = 3L/4, where L is the length of the connecting rod)

It was noticed that the calculated (theoretical) value of the transverse-vertical vibration accelerations increased with driving rotation speed or frequency increasing and the highest values were recorded when the point P was located at a distance of L/4 from the end A of the linear elastic connecting rod.

It may be noticed that the calculated (theoretical) values of the vibration acceleration on the vertical direction increased with an increasing action frequency or speed of the driving link and the highest values were recorded in the point placed at 1/4 of the length of the connecting rod measured from the end of action for lower speeds and in the point placed in the middle of the connecting rod for higher speeds.

In order to compare the vibrations displacements of the linear elastic links with the ones of linear viscoelastic links, there were considered three speeds of driving link for which it was computed the vibrations transversal displacements for both connecting rods, the one made of steel and the viscoelastic one made of un-masticated polyvinyl chloride, respectively a minimum speed, an average speed and a maximum speed of the driving link. There were not computed the vibrations longitudinal displacements because their values are insignificant (of the order of 10^{-10} ÷ 10^{-7}). For both connecting rods, the effective values of the displacements in the point placed in their middle u_2 (L/2, t) were computed for their comparison. It may be noticed that the computed (theoretical) values of the vibrations transversal displacements were greater for the viscoelastic connecting rod made of un-masticated polyvinyl chloride, PVC-U.

If, however, there are required for vibrations transversal displacements in the case of using the linear viscoelastic materials which does not exceed those of linear elastic metallic materials, this may be obtained increasing the cross section dimensions of the viscoelastic connecting rod. The materials with viscoelastic behavior are more advantageous in terms of comparable stiffness, both in terms of cost and because they have specific masses considerably lower in comparison with metallic materials, which is why the forces and moments of inertia are smaller.

In Fig. 1.6 the transverse-vertical acceleration variations were represented depending on the frequency of rotation of the driving element, for the connecting rod made of PVC-U, considering the following three particular positions of the point P: AP = L/4, AP = L/2 and AP = 3L/4, where L is the length of the connecting rod. It was noticed that the calculated (theoretical) values of the vibration acceleration in the vertical direction increased with speed or driving frequency increasing and the highest values were recorded when AP = L/4, at low rotation speed and in the middle of the rod at higher speeds.

The determination of stress and strains states due to free vibrations of the connecting rod of the mechanism was realized with the relations for the first approximation for the connecting rod made of steel and then for the connecting rod made of un-masticated polyvinyl chloride. The elements of the strains and stress tensors are computed with Mathematica program using the vibrations longitudinal displacements $u_1(x,t)$ and the vibrations transversal displacements $u_2(x,t)$ in the first approximation and their variation diagrams corresponding to the three speeds of the driving link chosen in the previous paragraph are represented.

The variation in time diagrams for the elements of the strains and stress tensors were also represented for the point placed at 1/4 of the length of the connecting rod from the end of action. It can be noticed that the elements of the two tensors have an increasing evolution with the increasing of the driving link speed for the both connecting rods. It can be also noticed that the elements of the two tensors record minimum values in the middle of the connecting rods, excepting the component ε_{11} whose values are maximum in this area(not significant, of the order of 10^{-8}).

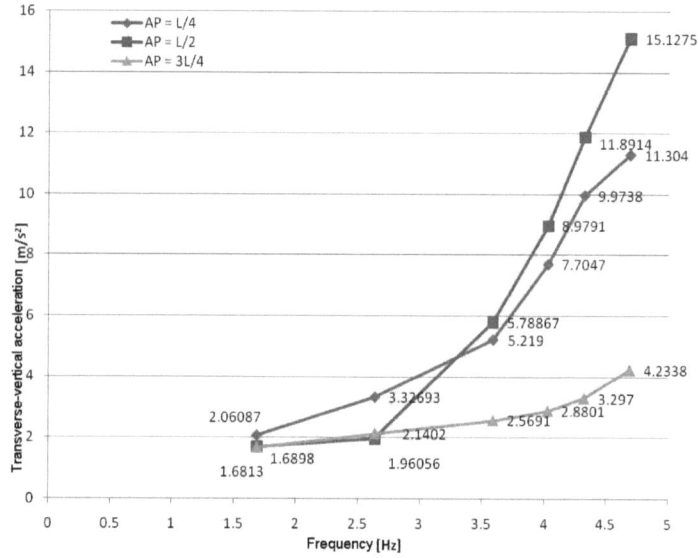

Fig. 1.6 Variation of cross-vertical acceleration of point P on the connecting rod made of PVC-U, depending on the driving frequency for the three considered positions of the point P (AP = L / 4, AP = L / 2 and AP = 3L / 4, where L is the length of the connecting rod)

In Fig. 1.7 there are represented the variations of the transverse-vertical displacement depending on the frequency of vibration of the driving link, in the point P placed in the middle of the connecting rod; it was noticed that the calculated (theoretical) transverse-vertical displacement of vibration was greater for the viscoelastic connecting rod made of PVC-U.

There were also calculated and there were represented the diagrams for the time variation of the strain and stress tensors elements in the point P located at the distance equal to L/4 from the end A of the connecting rod. So, for the driving link rotation speed n = 105.5 rpm (driving frequency f=1.758 Hz, angular speed ω_0 = 11.045 rad/s) there were obtained the time variation diagrams of the strain tensor elements $\varepsilon_{ij}(x,t)$ and $\varepsilon_{ij}(L/4,t)$, i, j = 1, 2, represented by Fig. 1.8 a) ÷ f).

The time variation diagrams of the stress tensor elements $\sigma_{ij}(x,t)$ and $\sigma_{ij}(L/4,t)$, i, j = 1, 2, are represented by Fig. 1.9 a) ÷ f).

Table 1.2 Effective values of the transverse-vertical displacements $u_2^{(1)}(L/2,t)$

Frequency [Hz]	1.758	3.662	4.321
$u_2^{(1)}(L/2,t)$[m] -	0.00073	0.00092	0.00136
$u_2^{(1)}(L/2,t)$[m] -	0.00319	0.00448	0.00869

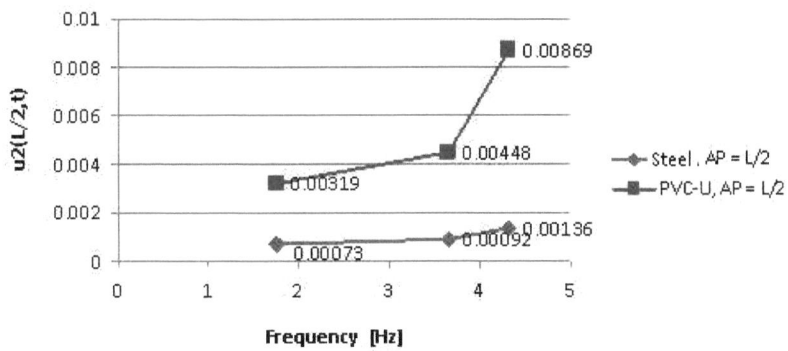

Fig. 1.7 Variation of transverse-vertical displacement of the point P placed in the middle of the connecting rod, depending on the driving frequency for the connecting rod made of steel rod and the connecting rod made of PVC-U

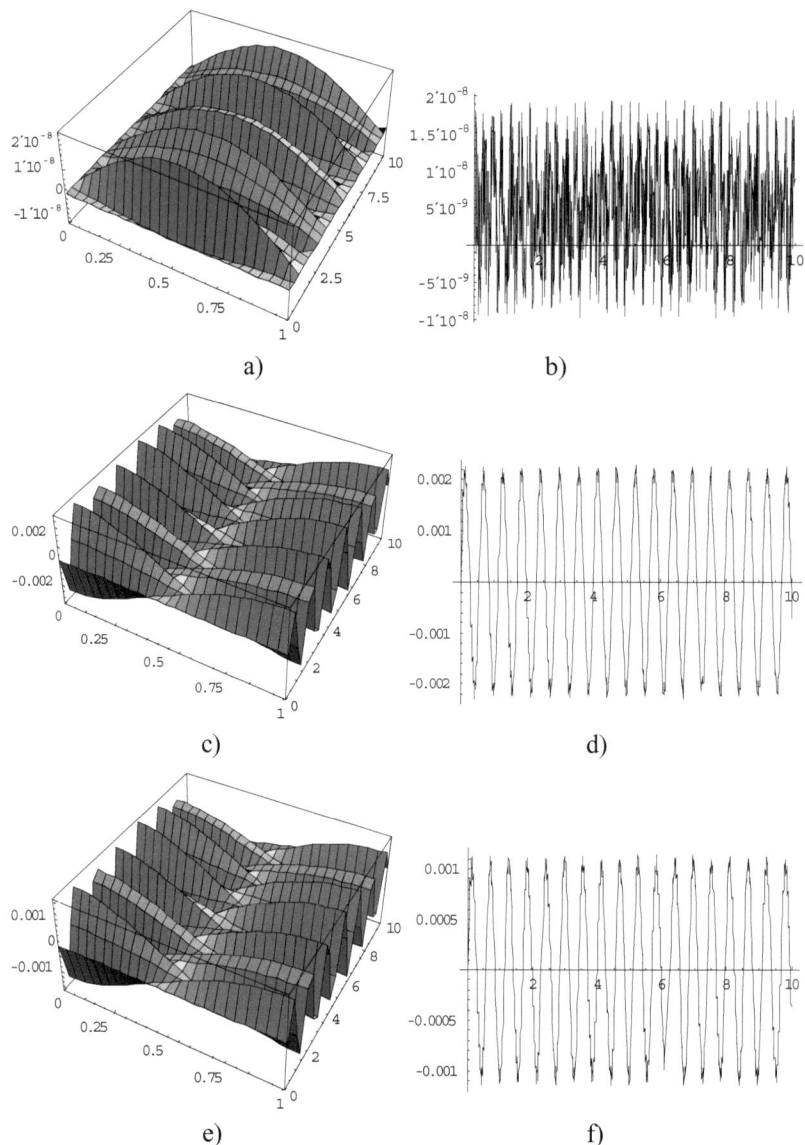

Fig. 1.8 The time variation diagrams of the strain tensor elements $\varepsilon_{ij}(x,t)$ and $\varepsilon_{ij}(L/4,t)$, $i, j = 1, 2$ for the connecting rod made of steel and the driving rotation speed n = 105.5 rpm: a) $\varepsilon_{11}^{(1)}(x,t)$; b) $\varepsilon_{11}^{(1)}(L/4,t)$; c) $\varepsilon_{22}^{(1)}(x,t)$; d) $\varepsilon_{22}^{(1)}(L/4,t)$; e) $\varepsilon_{12}^{(1)}(x,t)$; f) $\varepsilon_{12}^{(1)}(L/4,t)$.

24

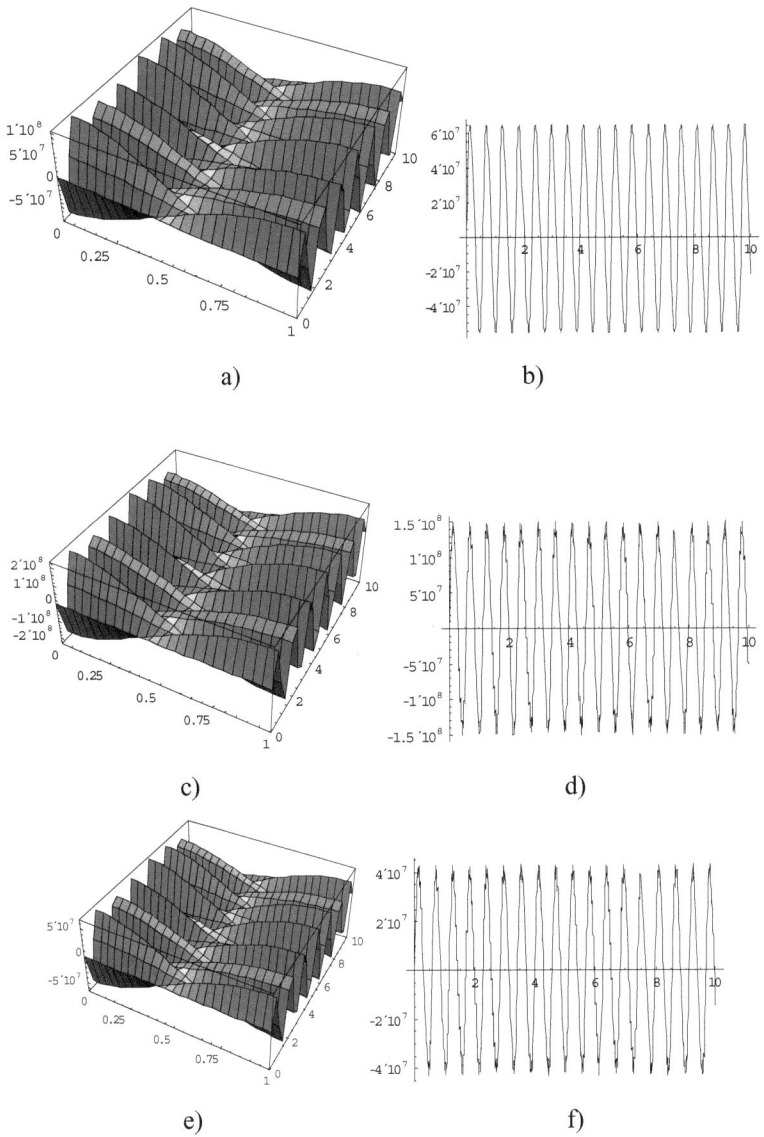

a)

b)

c)

d)

e)

f)

Fig. 1.9 The time variation diagrams of the stress tensor elements $\varepsilon_{ij}(x,t)$ and $\varepsilon_{ij}(L/4,t)$, $i,j = 1, 2$ for the connecting rod made of steel and the driving rotation speed n = 105.5 rpm: a) $\sigma_{11}^{(1)}(x,t)$; b) $\sigma_{11}^{(1)}(L/4,t)$; c) $\sigma_{22}^{(1)}(x,t)$; d) $\sigma_{22}^{(1)}(L/4,t)$; e) $\sigma_{12}^{(1)}(x,t)$; f) $\sigma_{12}^{(1)}(L/4,t)$.

It was noticed that the elements of the two tensors had an increasing trend, when the driving link rotation speed increased. Also, it was noticed that the elements of the two tensors registered minimum values in the middle of the connecting rod length except the component ε_{11} which, in this area, had maximum values (not significant, of 10^{-8} order). In order to compare the stress and strains states of the connecting rod with linear elastic behavior, subjected to free vibrations, with the ones of the viscoelastic connecting rod, there were first determined the effective values of the elements of the two tensors whose variation diagrams were presented in the previous paragraphs corresponding to the point placed at 1/4 of the length of the connecting rods from the end of action. The equivalent stress σ_{ech} was computed using the strength of materials fifth theory, the recommended theory for the elasto-plastic domain, using the effective values of the main stresses σ_1 and σ_2. Based on linear elastic longitudinal and transverse displacements, $u_1(x,t)$ and, respectively, $u_2(x,t)$ in the first approximation, determined for the connecting rod made of PVC-U the strain and stress tensors elements were determined using Mathematica software and there were represented the time variation diagrams corresponding to the chosen rotation speed values. There were also calculated the strain and stress tensors elements in the point P placed at L/4 from the end A of the connecting rod, where L was the length of the connecting rod and there were drawn their time variation diagrams. So, for the driving link rotation speed n = 105.5 rpm (driving frequency f = 1.758 Hz, angular velocity ω_0 = 11,045 rad / s) there were obtained the time variation diagrams of strain tensor elements represented by Fig. 1.10 a) ÷ f). The time variation of stress tensor components $\sigma_{ij}(x, t)]$ and σ_{ij} (L / 4, t), i, j = 1, 2, were shown in Fig. 1.11 a) ÷ f).

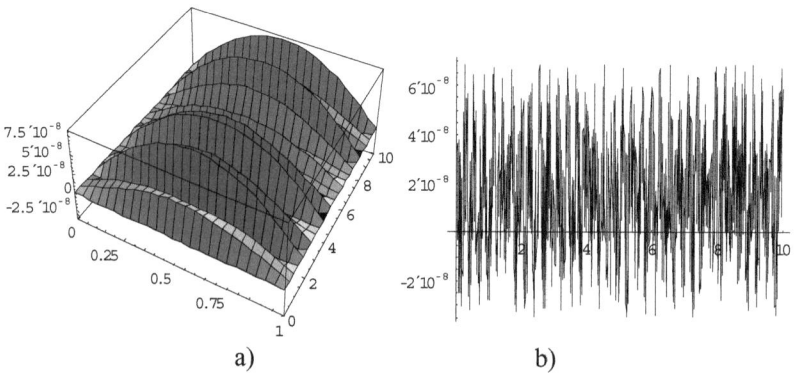

a)   b)

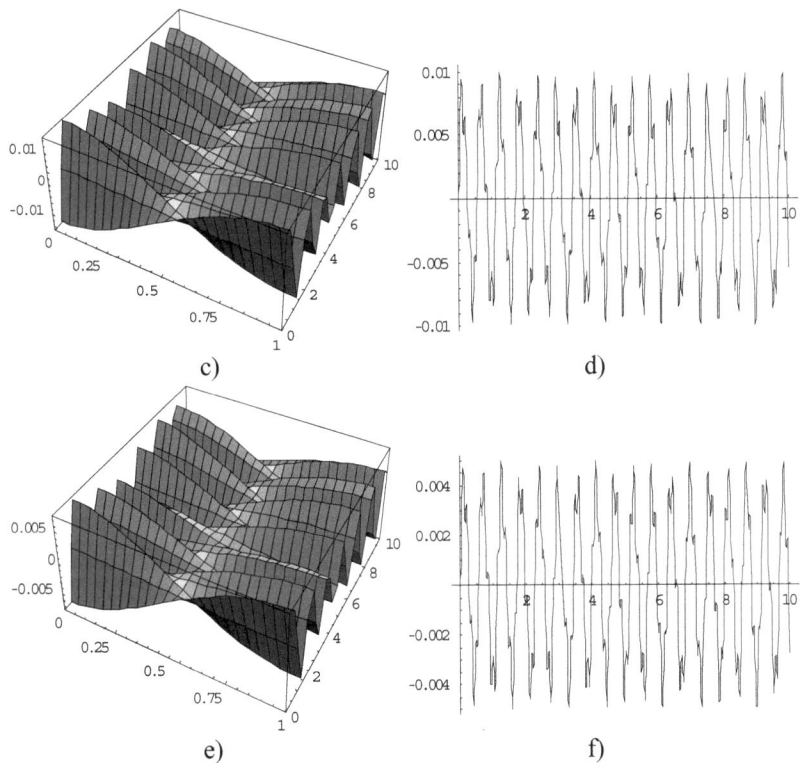

c)
d)

e)
f)

Fig. 1.10 The time variation diagrams of the strain tensor elements for the connecting rod made of PVC-U and the driving link rotation speed n = 105.5 rpm:

a) $\varepsilon_{11}^{(1)}(x,t)$; b) $\varepsilon_{11}^{(1)}(L/4,t)$; c) $\varepsilon_{22}^{(1)}(x,t)$; d) $\varepsilon_{22}^{(1)}(L/4,t)$; e) $\varepsilon_{12}^{(1)}(x,t)$; f) $\varepsilon_{12}^{(1)}(L/4,t)$.

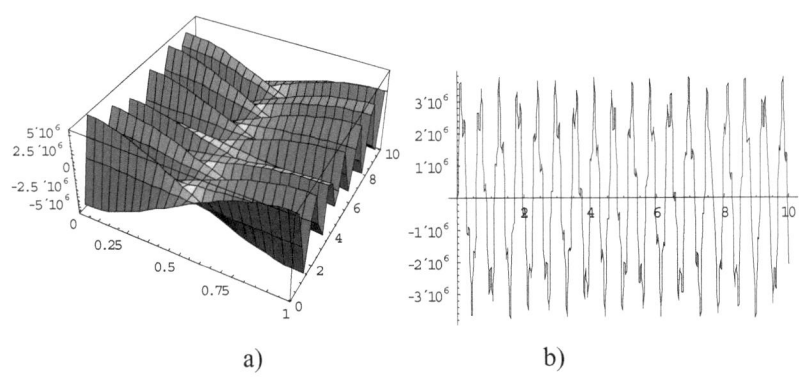

a)
b)

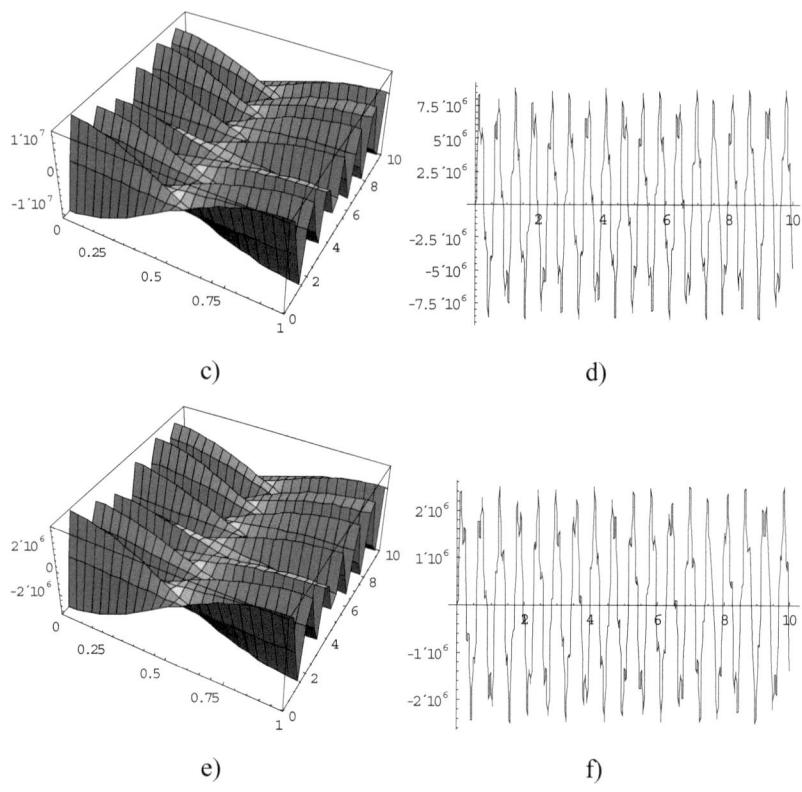

c) d)

e) f)

Fig. 1.11 The time variation diagrams of the stress tensor elements for the connecting rod made of PVC-U and the driving link rotation speed n = 105.5 rpm: a) $\sigma_{11}^{(1)}(x,t)$; b) $\sigma_{11}^{(1)}(L/4,t)$; c) $\sigma_{22}^{(1)}(x,t)$; d) $\sigma_{22}^{(1)}(L/4,t)$; e) $\sigma_{12}^{(1)}(x,t)$; f) $\sigma_{12}^{(1)}(L/4,t)$.

Analyzing the above presented time variation diagrams, it was noticed that also in the case of connecting rod made of PVC-U the two tensors elements had increasing trend, when the driving link rotation speed increased, and registered minimum values at the mid-length of the connecting rod; the exception was the component ε_{11}, which had maximum values in this area (not significant, its size order is about 10^{-8}). Thus there were obtained the data presented in Table 1.3, calculated for the specified drive frequencies of the driving link.

Figure 1.12 a) - c) shows the variations of the deformation tensors elements calculated for the point P placed at L / 4 distance from the end of the connecting rod, depending on the driving frequency, for the connecting rod made of steel, against the connecting rod made of PVC-U.

Table 1.3 The effective values of the strain and stress tensors components

No.	Tensors element	Frequency Connecting rod material	1.758 Hz	3.662 Hz	4.321 Hz
1.	ε_{11}	Steel	1.37442·	1.56682·1	2.17224
2.		PVC-U	4.83454·	7.49095·1	10.5608·
3.	ε_{22}	Steel	0.00160	0.002041	0.00301
4.		PVC-U	0.00702	0.023475	0.02358
5.	ε_{12}	Steel	0.00078	0.001016	0.00149
6.		PVC-U	0.00351	0.011359	0.01178
7.	$\sigma_{11}[N/m^2]$	Steel	0.46072·	0.587609·	0.87371·
8.		PVC-U	0.26682·	0.85624·1	0.88542·
9.	$\sigma_{22}[N/m^2]$	Steel	1.07523·	1.35849·1	1.98809·
10.		PVC-U	0.60111·	1.78191·1	2.06598·
11.	$\sigma_{12}[N/m^2]$	Steel	0.30762·	0.3883·10	0.56872·
12.		PVC-U	0.17788·	0.61536·1	0.61968·
13.	$\sigma_{ech}[N/m^2]$	Steel	1.07559·	1.35826·1	1.98726·
14.		PVC-U	0.60586·	1.8758·10	2.09165·

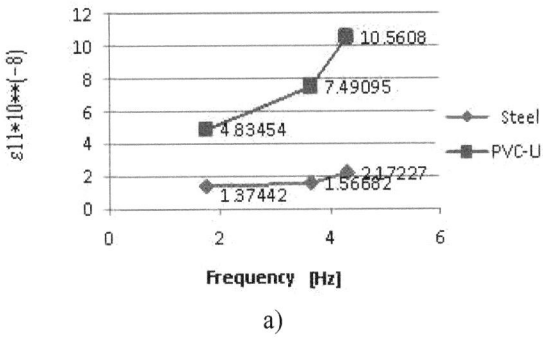

a)

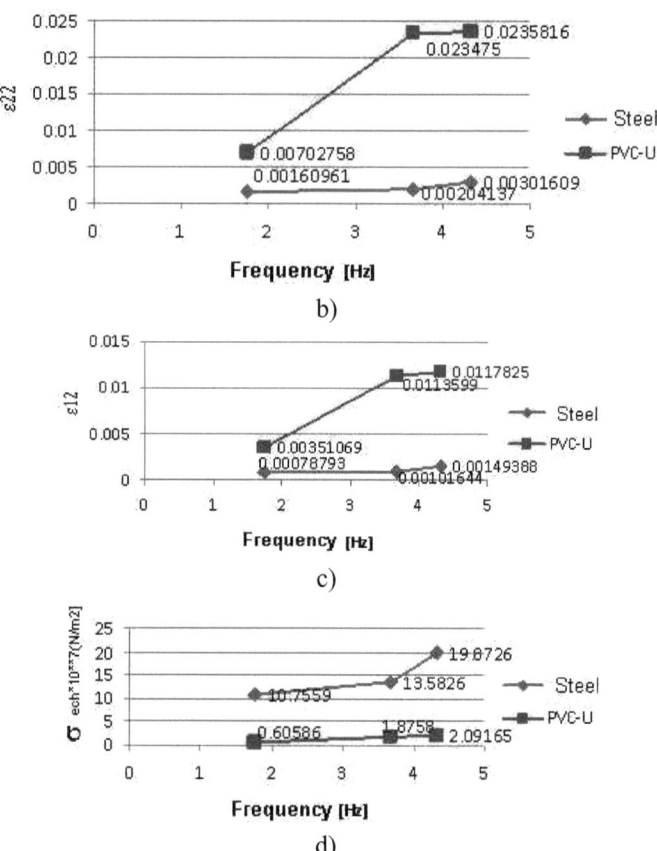

b)

c)

d)

Fig. 1.12 Variation diagrams of strain tensor elements and equivalent unit stress calculated for the point P placed at L / 4 distance from the end of the connecting rod, depending on the driving frequency, for the connecting rod made of steel, against the connecting rod made of PVC-U: a) ε_{11}; b) ε_{22}; c) ε_{12}; d) σ_{ech}.

Regarding the strains, it can be noticed that they are higher for the connecting rod made of PVC-U compared to that of steel, approximately 4 times greater at the lower action frequency and about 5-7 times greater at the highest action frequency, a fact that is explained by the lower stiffness of the viscoelastic material.

The elements of the stress tensor are higher for the connecting rod made of steel than for the one made of PVC-U. The equivalent stress is 16 - 17 times higher for the connecting rod made of steel than for the one made of PVC-U at the lower action frequency and about 9 times at the highest action frequency, due to the higher forces of inertia that appear in mechanism, when the connecting rod is made of a material with higher specific weight.

Concerning the modal analysis of the mechanism, the elements composing the crank and the connecting rod mechanism were first modeled in SolidWorks environment, then the assembly of the mechanism was modeled and it was performed the modal analysis using the module "visualNASTRAN INSIDE SolidWorks", application dedicated for numerical modeling with finite element in integrated designing. The module "visualNASTRAN INSIDE" associated to SolidWorks Program allows the use of finite element analysis with the latest capabilities of MSC/NASTRAN to automatically simulate the behavior of parts and assemblies modeled in SolidWorks. Modal analysis are presented in the thesis in appendices for the connecting rod made of OLC 45 and the one made of PVC-U, the operation frequency being f = 4.321 Hz.

Data presented by the modal analysis confirms the following conclusions:

- Concerning the motion of the mesh nodes, it can be remarked that the maximum value of these displacements is greater for the connecting rod made of PVC-U compared to that of steel, about 6 times at the highest action frequency, which is explained by the lower stiffness of the viscoelastic material;

- The equivalent stress is about 9 times higher for the connecting rod made of steel compared to that of PVC-U, due to the higher forces of inertia that appear in mechanism, when the connecting rod is made of a material with higher specific weight.

"The only source of knowledge is experience."
Albert Einstein

1.2.3 EXPERIMENTAL RESULTS

The purpose of the experiments was to check the influence of vibrations on the dynamics of R(RRT) mechanism and remark the differences between a connecting rod with a linear elastic behavior and a connecting rod with a linear viscoelastic behavior. It was used an experimental assembly and there were experimentally determined the vibrations transversal accelerations in three particular points placed on the both connecting rods.

The equipment used for the experimental determination of the dynamic response was composed of:

- Data acquisition system SPIDER 8;
- Load Amplifier Bruel & Kjaer 2635 type;
- Load Amplifier Robotron M1300 type;
- Accelerometers Bruel & Kjaer 4382 type;
- Inductive transducer for linear displacement WA300.

A mechanical system crank and connecting rod type was utilized, where it was mounted successively a 11x11x1000 connecting rod, made of two types of materials, steel and un-masticated polyvinyl chloride PVC-U (EN ISO 12608: 2003), and it was determined their vibrations response for variable action frequency between 1.5 and 5 Hz.

The tests were accomplished in the Machine Parts Laboratory of the Faculty of Mechanics from Craiova. An electric engine with a speed variator was used to operate and the driving was done with a trapezoidal belt.

The mechanism was set on a table with T-slots, placed on a rigid frame. The engine was fixed on this frame, too, in order to make the assembly as rigid as possible for eliminating other types of vibrations. The recordings were made with a sampling frequency of 4800 samples/second for during 35 ... 40 s.

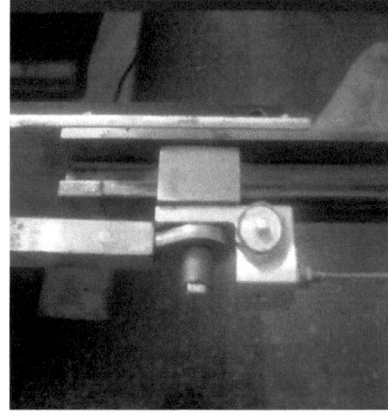

Fig. 1.13 Details for the accelerometers and stroke transducer mounting

It was made a "PrelExp" program under TestPoint programming environment for determining the vibrations response of the connecting rod in time and frequency.

Figure 1.14 shows the recorded original features for the connecting rod made of steel, for both accelerations in vertical and horizontal-transverse direction.

Relatively rigid behavior of the connecting rod causes vibrations to be progressively decreasing in vertical direction, starting from the driving end (the point placed at o quarter of the connecting rod length that had a higher vertical stroke) towards the end of translation.

Accelerations in horizontal-transverse direction are significantly lower than those in vertical direction.

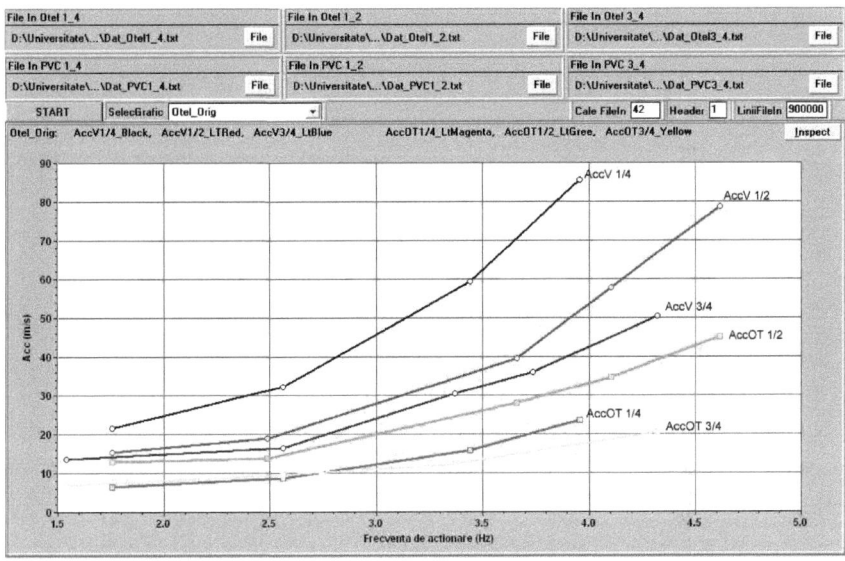

Fig 1.14 Original features for the connecting rod made of steel

Figure 1.15 shows original features recorded for the connecting rod made of PVC-U, for both accelerations in vertical and horizontal-transverse direction.

Flexible behavior of the connecting rod made the vibration response in vertical direction to depend on the driving frequency.

33

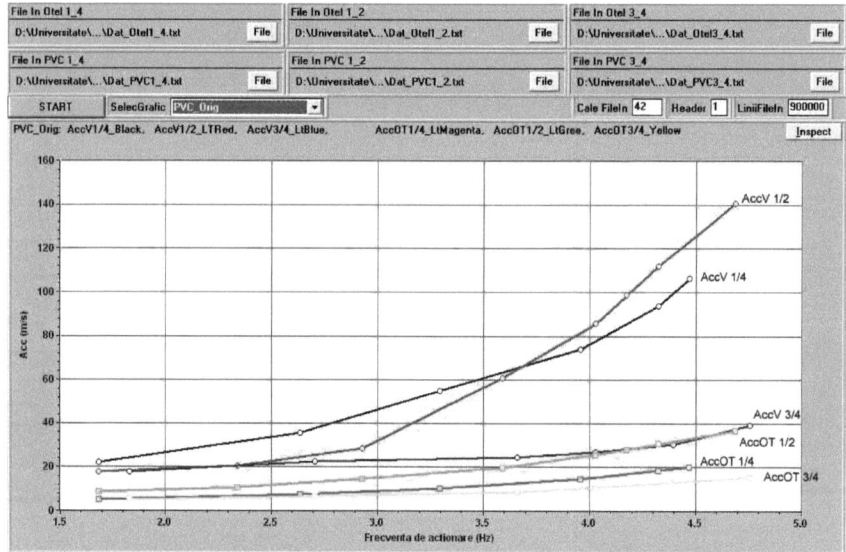

Fig. 1.15 Original features for the connecting rod made of PVC-U

"To expect the unexpected shows a thoroughly modern intellect."

Oscar Wilde

1.2.4 CONCLUSIONS

- Errors that occur between the theoretical values of the vibrations transverse acceleration obtained after solving the mathematical model and the experimental values are less than 9.3%.

- Spectral analysis showed the followings:

1. At the end of action, both connecting rods have a similar behavior. Spectral composition is similar. Small differences occur at higher action frequencies. Spectral components of high frequency occur for the connecting rod made of steel and the harmonica of the second order occurs, too. The spectral components of high frequency are reduced for the connecting rod made of PVC-U and, besides the fundamental, the first order harmonica occurs.

2. There are some differences between the two types of connecting rods in the point of measuring placed in their middle. The spectral components of superior frequency occur at the connecting rod made of steel and the harmonics of second and

34

fourth order are amplified. The spectral components of superior frequency are reduced at the connecting rod made of PVC-U and, besides the fundamental, the first order harmonica occurs with priority. Compared with the connecting rod made of steel, here the fundamental and the first order harmonica have greater amplitude, leading vibrations with higher effective value. For both connecting rods the transversal- horizontal oscillations respect the mentioned observations, but their amplitude is much smaller.

3. A relatively similar spectral composition occurs for both connecting rods in the point placed at 3/4 of the connecting rod length from the action end.

- It was observed that the elements of the stress and strains tensor have an increasing evolution when the driving element speed increase;

- Strains are higher for the connecting rod made of PVC-U, which is explained by the lower stiffness of the viscoelastic material;

- The elements of the stress tensor are considerably higher for the connecting rod made of steel, due to the higher forces of inertia that appear when the connecting rod is made of a material with higher specific weight.

- The paper underlines the influence of kinematic parameters on the vibrations displacements fields and implicit on the additional stress and strains states that occur due to vibrations in bar-type links, made of elastic or viscoelastic material.

The improvement of the solving methods of the mathematical models for viscoelastic links will contribute to the determination of the modes of vibrations for linear viscoelastic bars and will allow the observation of their behavior in operation.

The displacements fields computing by increasing accuracy also for the viscoelastic links will be the support for future determinations of stress and strains states, useful in mechanisms designing. This will open new directions of research in dynamics of rheological solids.

2. HOW CAN LUBRICANTS CHANGE THE WAY OF THE BEHAVIOUR OF THE ELASTIC ELEMENTS AT HIGHER SPEEDS OF MOTION

2.1 INTRODUCTION

Mechanisms having as components links, translational pairs, gear and so on, are not rigid in reality, they are elastic and suffer deformations when they are subjected to strong static or dynamic forces. In low speed motions, if the static forces are not strong, it is not necessary to take into account the elastic deformations, but in applications with high speed mechanisms may be designed inoperable in reality because of the large fluctuations of inertial forces.

As a result of more stringent requirements concerning higher work speeds and position accuracy of some points of the cinematic elements it is more necessarily to take into consideration the influence of the lubricant from the cinematic pairs. The pressure fields from the lubricant have a big influence above the vibrations that occur during the motion.

Our research tried to prevent and control the apparition of the lubricant film-breaking phenomenon, having consequences in the gripping and the working accuracy for the high precision mechanisms especially for the robotic parts.

2.2 PROBLEM FORMULATION

The considered mechanism is a slider-crank and connecting-rod assembly presented in Fig. 2.1. The vibrations of one long elastic element subjected to various spin speed of the leading element were experimentally measured. The rectilinear pair on the elastic element was lubricated with two kinds of oils.

Previously, solving the motion's equations of the elastic cinematic elements was made by assumption that these were continuous medium with an infinite degrees of freedom, or they were discrete systems (using finite element method), or using the Lagrange's method from elastodynamics [12], [19].

The equations of motion were solved using Hamilton's principle and combined it with the Reynolds's lubrication equation.

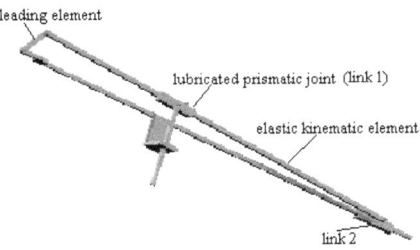

Fig. 2.1 The slider-crank and connecting-rod assembly

A crank and connecting-rod assembly with slide-bar was considered with rigid cinematic elements, excepting the 1m length element. A lubricated rectilinear pair with 0.100m length slides on this element.

Figure 2.2 shows the elastic cinematic element with the rectilinear pair.

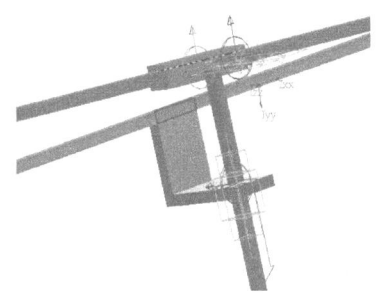

Fig. 2.2 Elastic cinematic element with mobile rectilinear pair and vertical element technologically loaded

2.3 ALGORITHM, EQUATIONS OF MOTION AND EQUATION OF THE PRESSURE FIELD

The long cinematic element with the rectilinear pair slides on was considered linear elastic, with plane motion. The equations of motion were obtained by using Hamilton's Principle from elastodynamics. The cinematic energy used in the application of this principle was given by generalized speeds field theory used for elastic bodies.

This fact supposes the directly inclusion of inertial terms in the equations of motion [2]. The pressures distribution was computed by using Reynolds equations of lubrication for a viscous and incompressible fluid. An averaging on the transverse direction (z) was applied to this fluid and it was computed depending on the external force practiced on the cinematic element by the fluid [17]:

$$\xi^3 \frac{\partial^2 P}{\partial X^2} + 3\xi^2 \frac{\partial \xi}{\partial X}\frac{\partial P}{\partial X} = 6\mu V_r(t)\frac{\partial \xi}{\partial X}, \tag{2.1}$$

where: P is the distributed pressure from the film, $v_r(t)$ is the relative speed between the elastic element and the mobile rectilinear pair, ξ is the width of the film, μ is the dynamic viscosity of the film.

More over, it was verified the next relation by using geometry:

$$\xi = \frac{H}{2} - u_2\left(x^* + X,t\right). \tag{2.2}$$

The boundary conditions for p(X,t) were given as:

$$X = 0 \Rightarrow P(0,t) = p_{atm}; \quad X = l \Rightarrow P(l,t) = p_{atm},$$

where : p is the atmospheric pressure.

The general solution for the equation (2.1) was:

$$P(X,t) = 6\mu V_r(t)I_1(X,t) + CI_2(X,t) + C_1, \tag{2.3}$$

where:

$$I_1(X,t) = \int dX \Big/ \left[\frac{H}{2} - u_2\left(X + x^*\right)\right]^2 \tag{2.4}$$

$$I_2(X,t) = \int dX \Big/ \left[\frac{H}{2} - u_2\left(X + x^*\right)\right]^3. \tag{2.5}$$

The constants C and C1 may be computed from the boundary conditions. The integrals I1 and I2 were computed by using Fourier Transform method.

Cavitation phenomenon occurs when the pressure computed with equation (2.3) decreases below the ambient pressure. In this case, Reynolds equation can't be applied inside the cavitation region.

The cavitation occurring position may be established by changing the boundary conditions (the pressures curve is easily translated in the cavitation region).

The boundary conditions are:

$$X = 0 \Rightarrow P(0,t) = p_{atm} ; \quad P(\overline{X},t) = p_{atm} \quad iar \quad \left.\frac{\partial p}{\partial X}\right|_{X=\overline{X}} = 0 \tag{2.6}$$

where: X is the position of the point where cavitation occurs.

The rectilinear pair was lubricated by oil with average lubricating properties and cinematic viscosity; its specific characteristics are shown in Table 2.1.

Table 2.1 The characteristics of the oils used as lubricants in the rectilinear pear

Oil	Density [kg/m^3]	Dynamic Viscosity [Pa·s]
TB32E without additives STAS 742/81	890	0.02848
SHELL TONNA T STAS 871/68	894	0.19668

Algorithm:

The solving algorithm steps were as follows:

- the cinematic and kinetostatic analysis of the mechanism was achieved considering the hypothesis of the rigidity of all its cinematic elements;

- the equations of motion were solved by introducing an external force R/(B*L) on the region occupied by the rectilinear pair and then, the pressures field was determined in the first approximation p1(X, t);

- the distributed force p1(X, t)/B, where B was the rectilinear pair width, was introduced in the new external force; then the equations of motion were solved again and so it was determined the pressures field in the second approximation.

This proceeding may continue by successively improving of the mechanism dynamic reply, depending on the wished accuracy of the calculus [19].

It was achieved a MAPLE Program for computing this algorithm and it was determined the necessary datum by two iterations, with enough accuracy, compared afterwards with the experimental data.

The mathematical model was solved by creating a program in Maple which two iterations MAPLE reached sufficient accuracy in two iterations and determined the bar-type link deformation and the pressure field in the lubricant.

The program has a high degree of generalization, it may be applied to any double articulated link in plane motion. The input variables of the program are the following:

- material features of the elastic kinematic element;
- dimensions of the elastic kinematic element (B * H * L);
- the length of the translational pair;
- viscosity properties of the lubricant;
- rotation speed of the driving element;
- data obtained from kinematic analysis of the mechanism (the reaction acting in the pair and the speed of the translational pair relative to the elastic element).

Note: For a different position of the elastic link in the kinematic chain new relationships may easily be adopted for a new calculation of speed and acceleration of the link end.

An important average used for simplifying the program was no considering the pressures field depending on time after the first iteration, but considering its effective value. Otherwise the model is difficult and it couldn't be solved anymore.

Furthermore, the pressure field from the rectilinear pair was modeled by finite element method with COSMOSM program. The specific heights of the oil film to the entrance and the exit of the pair were taken from the theoretical results of the deformations after the MAPLE program calculation of the deformations field.

2.4 PROBLEM SOLUTION

2.4.1 THEORETICAL RESULTS

The theoretical results for the link deformation field for the inferior oil used in the translational pair and different rotation speeds of the driving link are presented in Fig. 2.3, 2.4, 2.5, 2.6. In Fig. 2.7, 2.8, 2.9, 2.10 there are represented the link deformation field for the superior oil used in the translational pair, in the same work conditions.

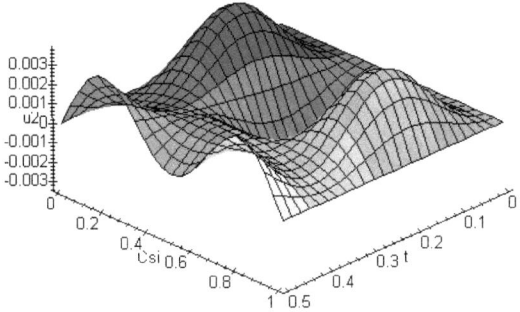

Fig. 2.3 The link deformations field for the 120 rpm (inferior oil)

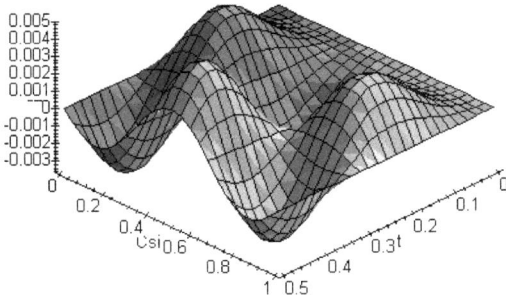

Fig. 2.4 The link deformations field for the 200 rpm (inferior oil)

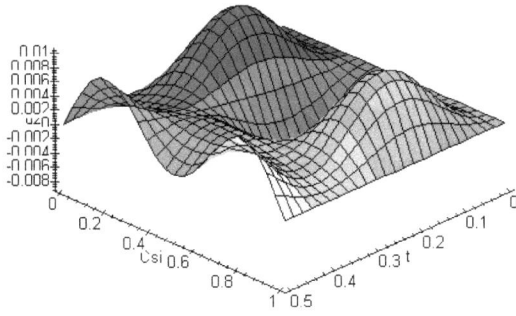

Fig. 2.5 The link deformations field for the 300 rpm (inferior oil)

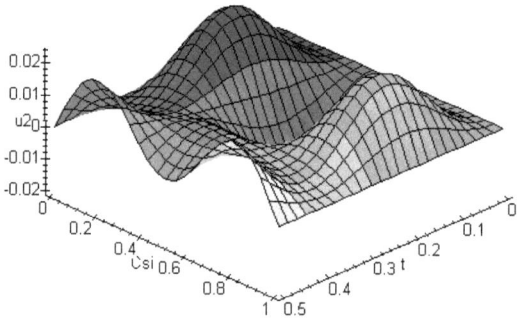

Fig. 2.6 The deformations field for the 387 rpm (inferior oil)

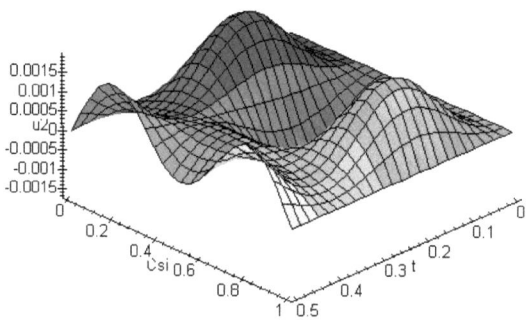

Fig. 2.7 The deformations field for the 120 rpm (superior oil)

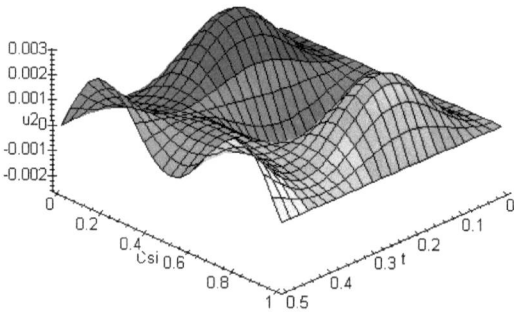

Fig. 2.8 The deformations field for the 200 rpm (superior oil)

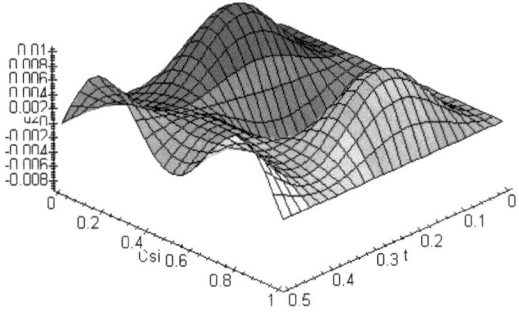

Fig. 2.9 The deformations field for the 300 rpm (superior oil)

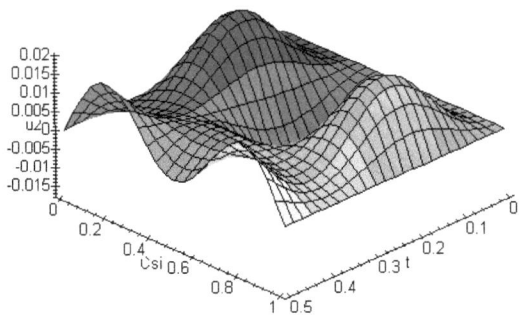

Fig. 2.10 The link deformations field for the 387 rpm (superior oil)

The theoretical pressure fields computed by the mathematical model for the two oils and different rotation speeds are represented in the following pictures.

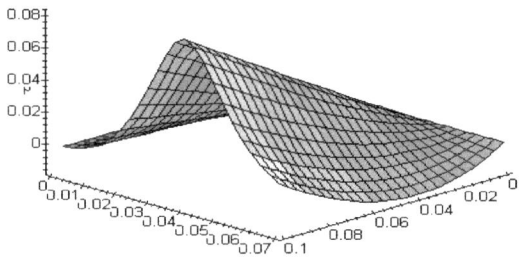

Fig. 2.11 Pressures field in lubricant at 120 rpm - superior oil

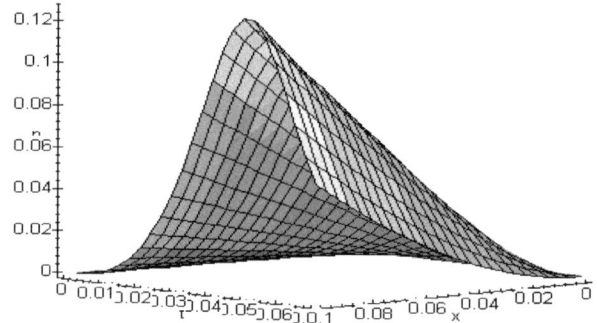

Fig. 2.12 Pressures field in lubricant at 200 rpm - superior oil

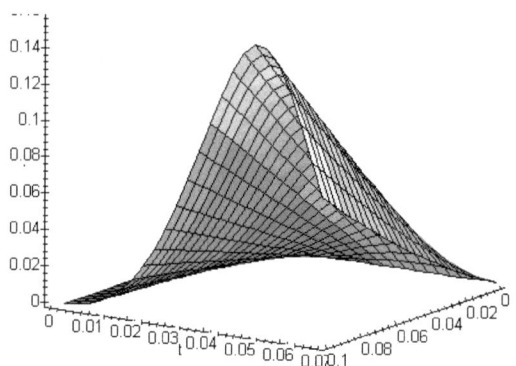

Fig. 2.13 Pressures field in lubricant at 300 rpm - superior oil

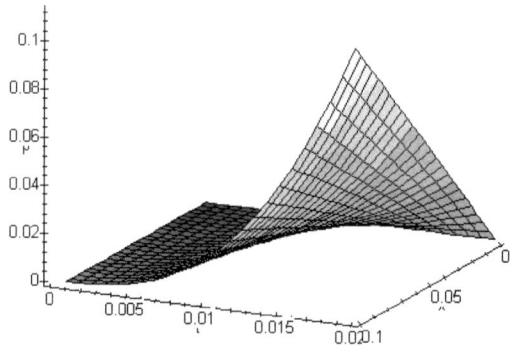

Fig. 2.14 Pressures field in lubricant at 387 rpm - superior oil

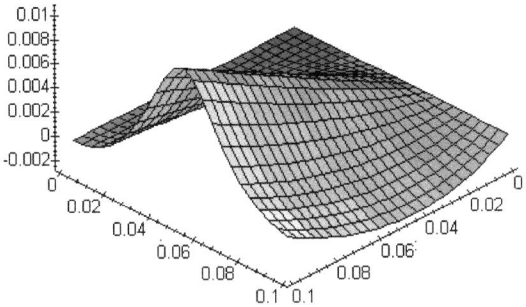

Fig. 2.15 Pressures field in lubricant at 120 rpm - inferior oil

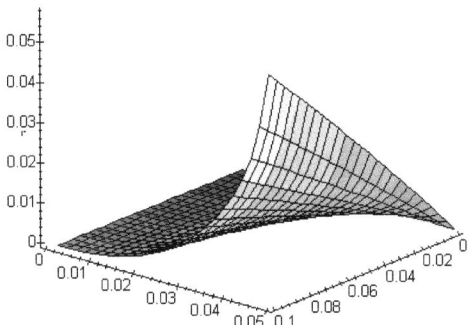

Fig. 2.16 Pressures field in lubricant at 200 rpm - inferior oil

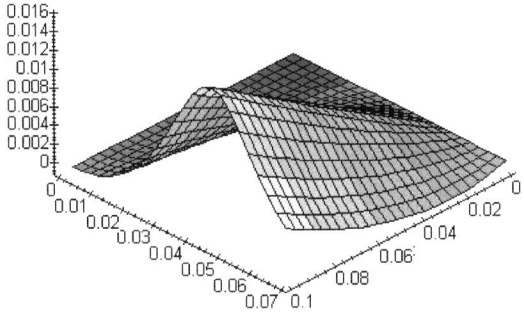

Fig. 2.17 Pressures field in lubricant at 300 rpm - inferior oil

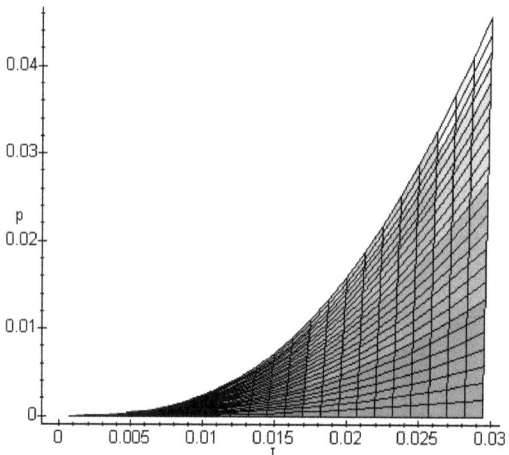

Fig. 2.18 Pressures field in lubricant at 387 rpm - inferior oil

"Experience is the teacher of all things."
Julius Caesar

2.4.2 EXPERIMENTAL RESULTS

The purpose of the experiments was to verify the influence of oil quality on the dynamics of parts in relative motion. The experiment design was based on the following reasons:

• Between parts in relative motion an oil film is forming, whose thickness depends on the quality of the used oil. From the point of view of vibration, the oil film is an elastic medium, being characterized by a modulus of elasticity which depends on the type of used oil and the thickness of the oil film. The assembly consisting of parts in relative motion - oil film represents an oscillating system (SO) with one degree of freedom, characterized by a certain characteristic frequency Fop.

• When external stimulus is applied on the moving part of an oscillating system, such as a sinusoidal motion with F_0 frequency, it gets a forced vibratory motion whose amplitude depends on the relative frequency F_0/F_{0p}. Vibration amplitude will reach its peak at resonance, when $F_0/F_{0p} = 1$. This is a periodic vibratory motion, with the same frequency as the external excitation.

• When a random excitation is applied to an oscillating system, a case encountered in practice, the forced motion will still be a random motion. Real systems may be modeled only aproximately as systems with one degree of freedom. In this case, the finite length of the mobile part and the finite thickness of the oil film has as effect the appearance of other degrees of freedom. This fact corresponds to higher order eigenfrequencies occurrence. In this case, random excitation may come from roughness of sliding path or impurities on it, temporary loss of oil film etc. In this case, random excitation with frequency band ΔF will excite the degrees of freedom located in this band. The disrupted motion may be such amplified as to lead to oil film rupture and system crashes, depending on the oil quality and sinusoidal oscillation frequency.

It has been achieved an experimental assembly in order to highlight these issues. There were determined the followings:

• linear stroke of the translational pair 2;

• vibration transverse acceleration of the main lubricated slide motion.

In Fig. 2.19 it may be noticed how the lubricant film was broken (dark areas on kinematic elastic element). The mechanism was disassembled before taking the picture in order to catch the entire stroke of the lubricated translational pair. It may be seen how lubricant film breaking led to the deterioration of the rectified zone of the elastic kinematic link determining the longitudinal stripes appearance.

The experimental results using the accelerometer device give the effective values of the vibration acceleration and the spectral analysis for every kind of oil and different values of the driving element rotation speed.

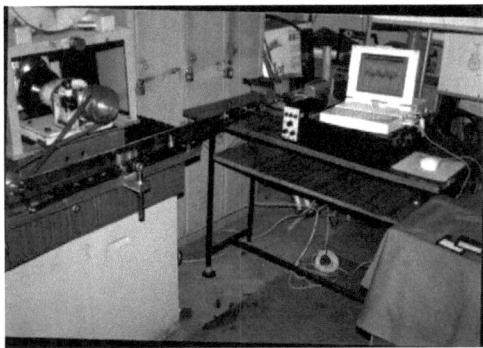

Fig. 2.19 Experiments equipment

47

Fig. 2.20 Damaged elastic kinematic link
because of breaking oil film

Figure 2.21 and 2.22 present the same experimental determinations for the inferior and the superior oil respectively. The effective value of the vibration acceleration is 106.56 m/s^2 for the inferior oil and 80.37 m/s^2 for the superior oil.

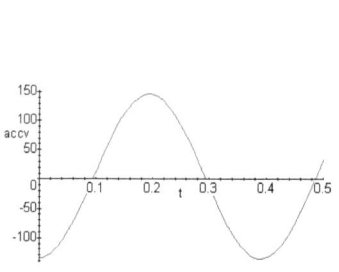

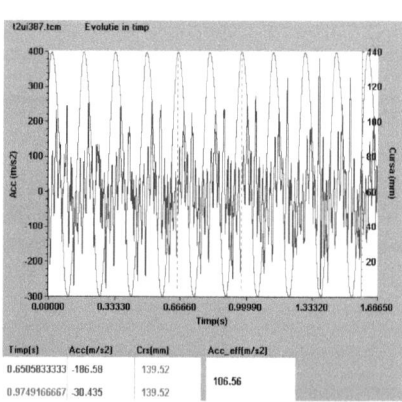

Fig. 2.21 The vibration acceleration for the inferior oil at 387 rpm

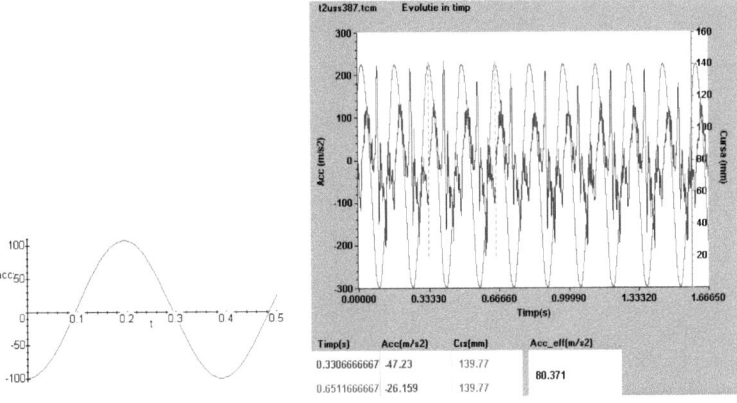

Fig. 2.22 The vibration acceleration for the superior oil at 387 rpm

The spectral analysis for the inferior oil and 387 rpm is presented in Fig. 2.23 and Fig. 2.24 presents the spectral analysis for the superior oil. The experimental results were compared with the ones given by the mathematical model solving. The compared characteristic quantity is the vibration acceleration and its values are shown in Table 2.2 for the inferior oil (TB32E).

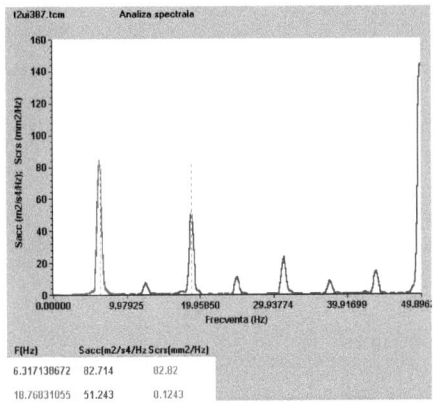

Fig. 2.23 The spectral analysis for the inferior oil at 387 rpm

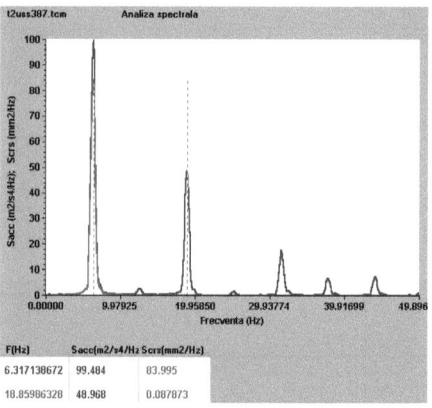

Fig. 2.24 The spectral analysis for the inferior oil at 387 rpm

Table 2.2 Comparison between the theoretical vibration acceleration for the inferior oil at 387 rpm

Rotation [rpm]	120	200	300	387
Theoretical Vibration Acceleration [m/s^2]	7.23	13.47	62.41	106.06
Experimental Vibration Acceleration [m/s^2]	7.66	13.301	63.61	106.56

The compared characteristic quantity is the vibration acceleration and its values are shown in Table 2.3 for the superior oil (SHELL TONNA T).

Table 2.3 Comparison between the theoretical vibration acceleration for the superior oil at 387 rpm

Rotation [rpm]	120	200	300	387
Theoretical Vibration Acceleration [m/s^2]	5.39	12.41	51.06	79.43
Experimental Vibration Acceleration [m/s^2]	5.59	12.732	51.44	80.37

The lubrificant film thicknesses at the entrance, respectively at the exit of the rectilinear pair, are computed with the datum obtained by solving the mathematical model. It was achieved a finite element routine COSMOS with this datum, considering the lubricant film as a plane laminar boundary layer.

A program in COSMOS programming environment was made in order to determine the lubricant pressure field. This program used as input the properties of the lubricant viscosity and density, the translational pair geometry and the link deformations at the entrance and, respectively, at the exit of the pair at a moment of time, data which were taken from the solutions of the mathematical model.

The obtained values were compared to those given by solving the mathematical model, comparison highlighted by Table 2.4.

Table 2.4 The pressure field values – theoretical and given by COSMOS

Oil type and speed	Theoretical value of distributed pressure field [N/m]	The pressure distribution given by COSMOS [N/mm]
Inferior Oil, 120 rpm	0.005	4.367
Inferior Oil, 200 rpm	0.0106	10.07
Inferior Oil, 300 rpm	0.0159	15.44
Inferior Oil, 387 rpm	0.016	16.008
Medium oil, 120 rpm	0.0081	8.347
Medium oil, 200 rpm	0.012	12.67
Medium oil, 300 rpm	0.0235	24.092
Medium oil, 387 rpm	0.0289	30.21
Superior oil, 120 rpm	0.044	44.08
Superior oil, 200 rpm	0.078	71.62
Superior oil, 300 rpm	0.096	95.8
Superior oil, 387 rpm	0.110	107.57

From the above data it was noticed that the errors that occurred during solving by the two methods were at most 8.12%.

The mechanism used for experimental research was made with a high degree of precision, the rotation pairs were provided with high precision bearings and there were taken precautions to eliminate additional clearances which might determine unwanted vibrations in mechanism assembly. The slide passes through a vertical ball bearing with helical channel and the elastic kinematic element zone participating to the translational pair stroke was rectified.

There were compared the theoretical and experimental values of the vibration acceleration shown in Table 2.5.

Table 2.5 Theoretical and experimental values of the vibration acceleration

Oil type and speed	The vibration acceleration value obtained theoretically [m/s^2]	The vibration acceleration value obtained experimentally [m/s^2]
Inferior Oil, 120 rpm	7.23	7.66
Inferior Oil, 200 rpm	13.47	13.301
Inferior Oil, 300 rpm	62.41	63.61
Inferior Oil, 387 rpm	106.06	106.56
Medium oil, 120 rpm	6.41	6.38
Medium oil, 200 rpm	12.76	13.144
Medium oil, 300 rpm	53.9	53.7
Medium oil, 387 rpm	87.9	86.34
Superior oil, 120 rpm	5.39	5.59
Superior oil, 200 rpm	12.41	12.732
Superior oil, 300 rpm	51.06	51.44
Superior oil, 387 rpm	79.43	80.37

As it may be noticed in Table 2.5, errors occurring between theoretical and experimental values were up to 5.61%. Also, as long as the driving element speed was small, the variances between the three types of oils were smaller. And when the driving element speed increased, the differences became larger.

Equally, the elastic link deformations were higher as the driving element speed was higher and the oil quality was inferior. Whet it was used inferior oil, the phenomenon of the lubricant film breaking appeared, which lead on to a malfunction of the mechanism and elastic link wear. In this case vibrations were even higher and the chosen mathematical model did not match.

"Failure is simply the opportunity to begin again, this time more intelligently."
Henry Ford

2.4.3 PROBLEMS...

Cavitation occurs when the film pressure falls below atmospheric pressure. This phenomenon was also observed in the COSMOS programming environment using finite element method. It occurred in the mathematical model, especially for high speeds, at certain points of time. In these areas and these points of time, the mathematical model did no longer correspond because the boundary conditions for the cavity phenomenon were different from those used in the case of the lubricant film operation.

The paper highlights the influence of the translational pair lubricant on the elastic kinematic element vibrations. It was noticed that as the oil was of inferior quality and the speeds were higher, vibrations appeared with larger amplitude.

As a result, the accuracy of the mechanism, i.e. the follower rod peak position varied from 0.005 to 0.006 m for inferior oil and from 0.0019 to 0.003 m for superior oil in the used range of the rotation speed. Moreover, for the inferior oils with inferior lubrication characteristics, the phenomenon of lubricant film breaking appears and leads to seizing and reducing of operating precision for high precision mechanisms.

The influence of lubricant from the translational pair on vibrations of the elastic kinematic element was determined experimentally, namely the amplitude of vibration for the lubricant oil with medium characteristics ranged from 0.0014 to 0.0038 m, leading to lower values of operation precision of mechanism.

In Fig. 2.25 - 2.28 there are presented the pressures distribution in inferior lubricant for different rotation speed values and in Fig. 2.29 - 2.32 there are presented the pressures distribution in superior lubricant for the same rotation speed values.

In Fig. 2.28 it may be noticed that the pressure value decreased below 0; this means that cavitation phenomenon appeared.

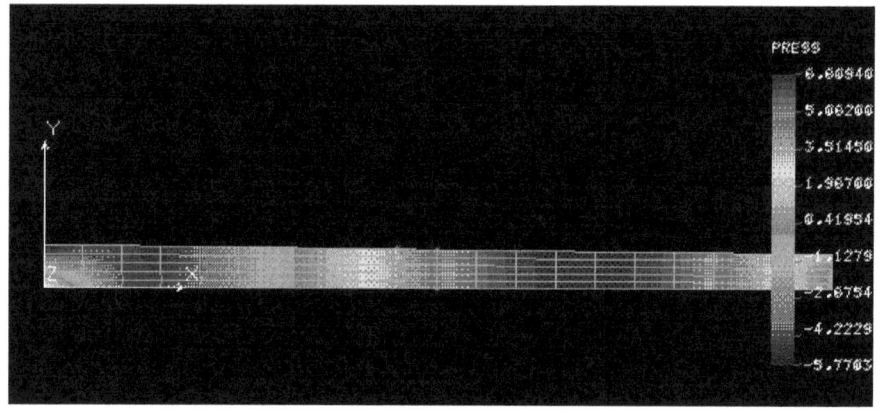

Fig. 2.25 Pressures distribution in inferior lubricant for 120 rpm rotation speed value

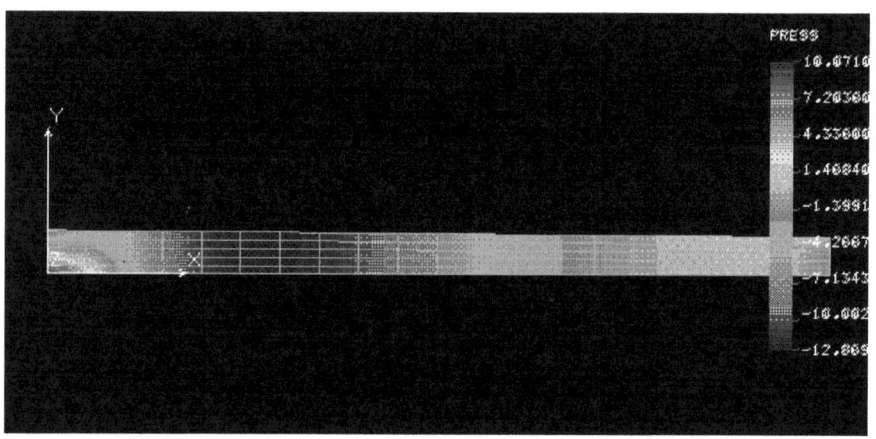

Fig. 2.26 Pressures distribution in inferior lubricant for 200 rpm rotation speed value

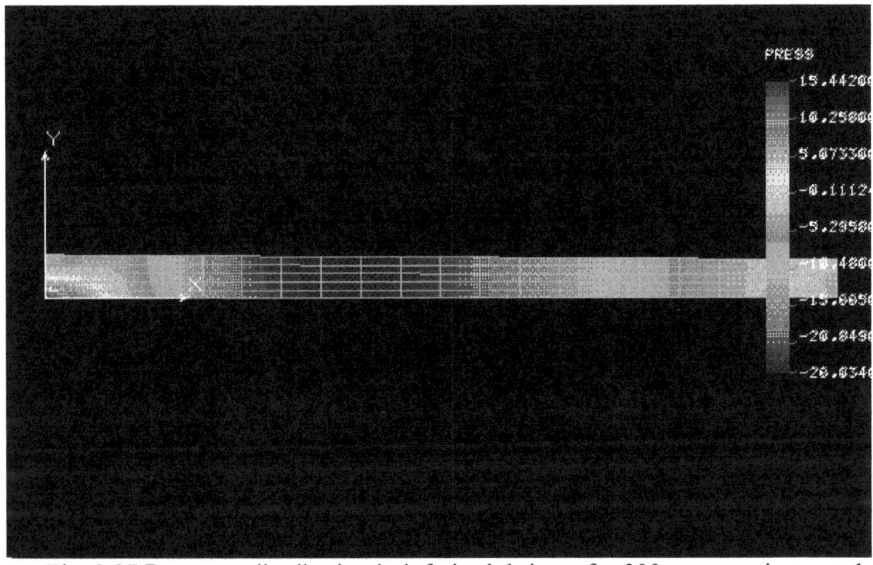

Fig. 2.27 Pressures distribution in inferior lubricant for 300 rpm rotation speed value

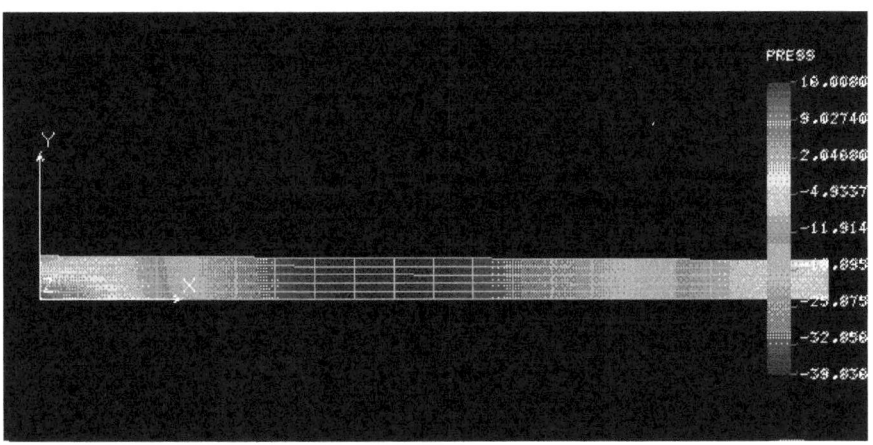

Fig. 2.28 Pressures distribution in inferior lubricant for 387 rpm rotation speed value

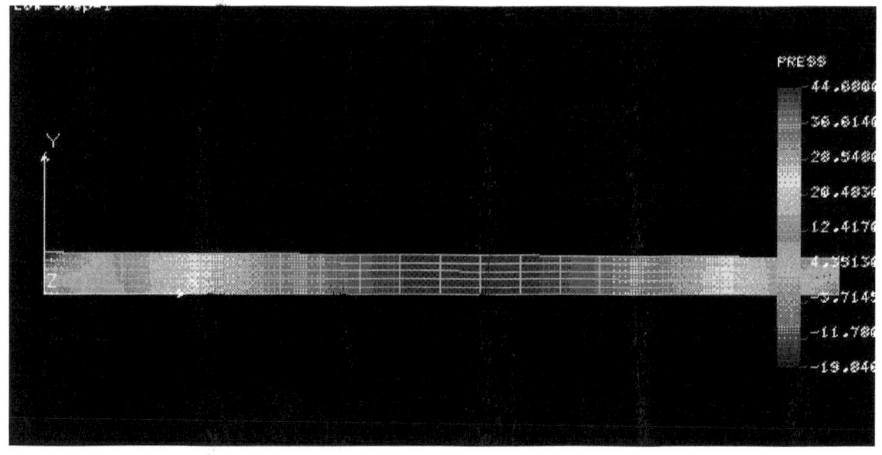

Fig. 2.29 Pressures distribution in superior lubricant for 120 rpm rotation speed value

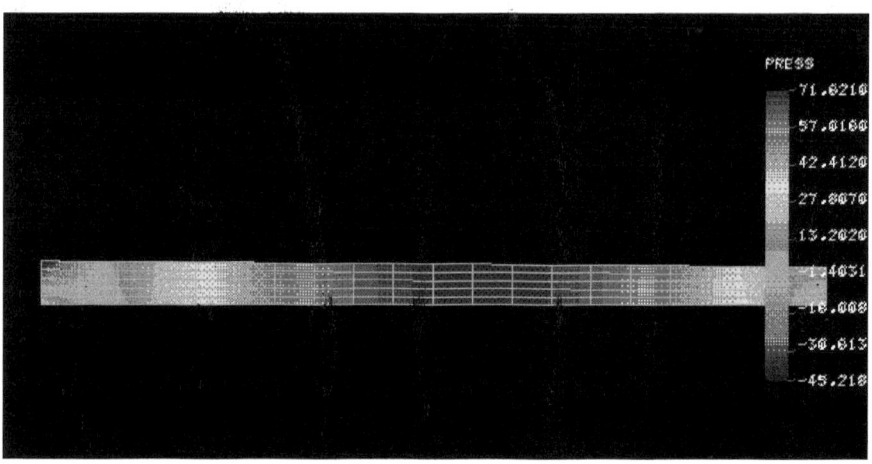

Fig. 2.30 Pressures distribution in superior lubricant for 200 rpm rotation speed value

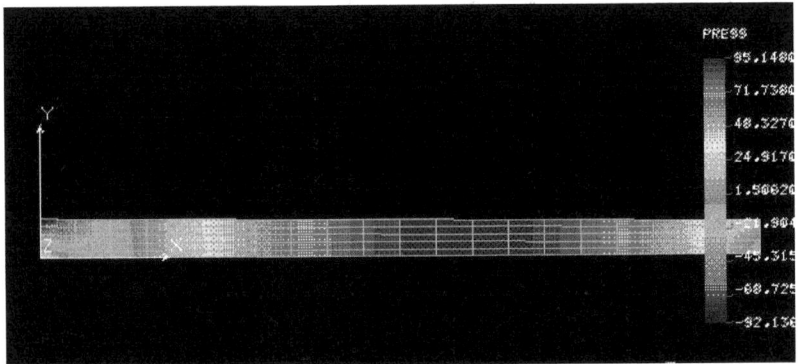

Fig. 2.31 Pressures distribution in superior lubricant for 300 rpm

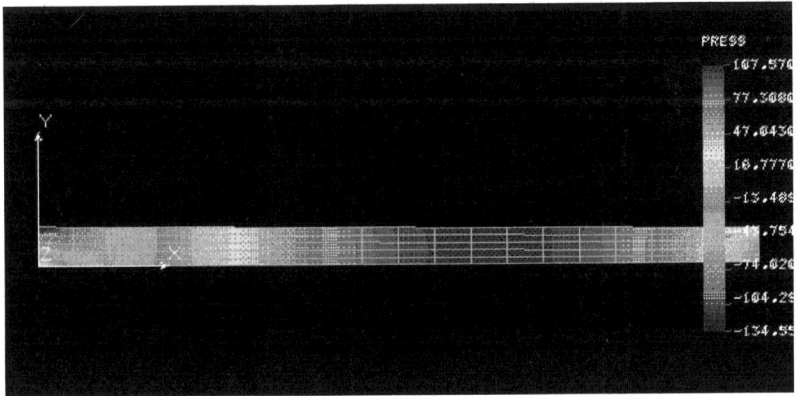

Fig. 2.32 Pressures distribution in superior lubricant for 387 rpm

The comparison of the theoretical and experimental results for the values of the pressures field is revealed in the Table 2.6.

Table 2.6 The comparison of the theoretical and experimental results for the values of the pressures field, at 387 rpm rotation speed value

Type of the lubricant	Theoretical value for the pressure in lubricant [N/m]	The COSMOSM value for the pressure in lubricant [N/m]
Inferior	0.0160	0.016008
Superior	0.1100	0.10757

"It is always wise to look ahead, but difficult to look further than you can see."

Winston Churchill

2.5 CONCLUSIONS

The influence of the lubricant from the rectilinear pair on the elastic cinematic element vibration was determined on the experimental way: the vibration amplitude varied from 0.005 m to 0.006 m for the oil with inferior characteristics and made decreasing the mechanism working accuracy. Besides, phenomenon like cavitation and lubricant film breaking appeared in the pressure field from the lubricant film and they may bring about gripping and deteriorating accurately mechanism's elements.

The vibration amplitude varied from 0.0019 m to 0.003 m for the superior oil and the cavitation phenomenon didn't appear for the used range of speeds.

MAPLE routine has a great level of generality. It may be applied to any plane-moving element with a mobile lubricated rectilinear pair. This routine computes the pressure field in lubricant and the maximal deformations of an elastic element with the accuracy less than 8%. It may be used as entrance variable quantity: material properties of the elastic elements, its dimensions, the lubricant viscosity characteristics and the length of the rectilinear pair, as well as the driving element rotation. The behavior of the mechanism and its precision in work is so better so the oil is better as quality: the properties of lubrication and viscosity. The same situation is obtained when the speed is low. Knowing all these characteristics the deterioration of the elements of the mechanism can be avoided.

In the same time, the differences between the oils are low when the speed is low. When the speed of the leader element is high the differences become higher.

And a final conclusion…

"A scientific truth does not triumph by convincing its opponents and making them see the light, but rather because its opponents eventually die and a new generation grows up that is familiar with it."

Max Planck

References:

1. Bagnaru, D.Gh., Vibratiile elementelor cinematice, Ed. SITECH, Craiova, 2005

2. Buculei, M., Marghitu, D., Contributii la studiul influentei filmului de lubrifiant din cuplele de translatie la analiza vibratiilor elementelor rectilinii ale mecanismelor, Simp. MTM 1980

3. Calbureanu, M., Lanturi cinematice cu elemente elastice si cuple de translatie lurifiate, Ed. Universitaria, Craiova, 2008

4. Calbureanu, M., Malciu, R., Bagnaru, D., Aspects about the vibrations of a linear viscoelastic cinematic element of crank and connecting rod assembly, S1, pag. 140-145, Annals of the Oradea University, Fascicle of Management and Technological Engineering, Vol. VII (XVII), ISSN 1583-0691, 2008

5. Calbureanu, M., Malciu, R., Lungu, M., Calbureanu, D., Aspects about the influence of the lubricant from a rectilinear pair above the work accuracy of the elastic elements from the high precision mechanisms, 12th WSEAS International. Confference CSCC EE EMESEG'08, pp. 209-214, ISBN 978-960-6766-88-6, ISSN 1790-2769, 2008

6. Calbureanu, M., Malciu, R., Lungu, M., Calbureanu, D., The influence of the lubricant from a rectilinear pair above the work accuracy of the elastic elements from the high precision mechanisms, WSEAS TRANSACTIONS on APPLIED and THEORETICAL MECHANICS, ISSN: 1991-8747, p.176-185, Issue 5, Volume 3, May 2008

7. Calbureanu, M., Malciu, R., Lungu, M., Dumitru, S., Analytic determination of the accelerations of vibrations for a linear viscoelastic cinematic element of a crank and connecting rod assembly validated by experiment, Proceedings of the 11-th WSEAS International Conference on Automatic Control, modeling and Simulation (ACMOS'09), Istanbul, Turkey, ISSN: 1790-5117, ISBN: 978-960-474-082-6, Published by WSEAS Press, pp. 349-354, 2009.

8. Cherciu, M., Dumitru, N., Computer Aided Kinematic Modelling of the Mobile Mechanical Systems, The Ninth IFToMM International

Symposium on Theory of Machines and Mechanisms, SYROM 2005, pp. 861, 2005

9.   Dumitru, N., Cherciu M., Zuhair, A., Theoretical and Experimental Modelling of the Dynamic Response of the Mechanisms With Deformable Kinematics Elements, IFToMM, Besancon, France, 2007

10.   Dumitru, N., Cherciu, M., Zuhair, A., The Finite Element Modeling Of The Dynamic Response Of The Mechanisms, -13th International Congress of Sound Vibration, Vienna, Austria, ISBN 3-9501554-5-7, 2006

11.   Dumitru, N., Malciu, R., Calbureanu, M., Dumitru, S., Marinescu, G. C., Dynamic Analysis of a Mobile Mechanical System with Deformable Elements - Advanced Materials Research (Volumes 463 - 464) Pg: 1242-1245, DOI: 10.4028 / www.scientific.net / AMR.463-464.1242

12.   Erdman, A.G., Sandor, G.N., A General Method for Kineto-Elastodynamics Analysis and Synthesis of Mechanisms, Journal of Engineering for Industry, Vol. 94, pp. 1193-1205,1972

13.   Kuo, Y.L., Cleghorn, W.L., Application of Stress-based Finite Element Method to a Flexible Slider Crank Mechanism, 12th IFToMM Congress, Besancon, June 18-21, 2007

14.   Kuo, Y.L., Cleghorn, W.L., Behdinan, K., Applications of Stress-based Finite Element Method on Euler-Bernoulli Beams, Proceedings of the 20th Canadian Congress of Applied Mechanics, Montreal, Quebec, Canada, May 30-Jun 2, 2005

15.   Kuo, Y.L., Cleghorn, W.L., Behdinan, K., Stress-based Finite Element Method for Euler-Bernoulli Beams, Transactions of the Canadian Society for Mechanical Engineering, Vol. 30(1), pp. 1- 6, 2006

16.   Malciu, R., Calbureanu, M., Lungu, M., Dumitru, S., Comparison between vibrations transversal displacements analytic determined for a linear elastic connecting rod and a linear viscoelastic one (after the validation by experiment of the vibrations effective accelerations), Proceedings of the 11-th WSEAS International Conference on Automatic Control, modeling and Simulation (ACMOS'09), Istanbul, Turkey, ISSN: 1790-5117, ISBN: 978-960-474-082-6, Published by WSEAS Press, pp. 355-360, 2009

17.   Oroveanu, T., Mecanica fluidelor vascoase, Ed. Academiei, Bucuresti, 1967.

18.   Swider, J., Wszolek, G., Michalski, P., The numerical software for vibration analysis of mechanical systems, 9-th Conference on Dynamical Systems, Theory and Applications, Lodz, Poland, pp. 943-950, December, 2007

19.   Zienkiewicz, O. C., Element The Finite Method in Engineering Science, McGraw Hill, 1971

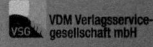

Printed by Books on Demand GmbH, Norderstedt / Germany